The Collected Economics Articles
of Harold Hotelling

The Collected Economics Articles of Harold Hotelling

Edited by and
with an Introduction by
Adrian C. Darnell

With 10 Illustrations

Springer-Verlag
New York Berlin Heidelberg
London Paris Tokyo

Adrian C. Darnell
The Department of Economics
University of Durham
Durham DH1 3LE
England

Mathematics Subject Classifications (1980): 90Axx, 90A10, 90A11, 90A12, 90A14, 90A15, 90A16

Library of Congress Cataloging-in-Publication Data
Hotelling, Harold, 1895–
  The collected economic articles of Harold Hotelling.
  Includes bibliographical references.
  1. Hotelling, Harold, 1895–   . 2. Economics,
Mathematical.   3. Mathematical Statistics.   I. Darnell,
A. C.   II. Title.
HB119.H68   1989   330'.01'51   89-21666
ISBN-13:978-1-4613-8907-1

Typeset by Asco Trade Typesetting Ltd., Hong Kong.

9 8 7 6 5 4 3 2 1

ISBN-13:978-1-4613-8907-1   e-ISBN-13:978-1-4613-8905-7
DOI: 10.1007/978-1-4613-8905-7

Preface

In 1985 I first began my research on the life and work of Harold Hotelling. That year, Harold Hotelling's widow had donated the collection of his private papers, correspondence and manuscripts to the Butler Library, Columbia University. This is a most appropriate place for them to reside, in that Hotelling's most productive period as an active researcher in economics and statistics coincides with the years when he was Professor of Mathematical Economics at Columbia (1931–1946). The Hotelling Collection comprises some 13,000 separate items and contains numerous unpublished letters and manuscripts of great importance to historians of economics and statistics.

In the course of the following year I was able, with the generous financial assistance of the Nuffield Foundation, the Economic and Social Research Council, the British Academy and the University of Durham, to spend six weeks over the Easter period working on the collection. I returned to New York in September 1986 while on sabbatical leave from the University of Durham, and I spent most of the following eight months examining the many documents in the collection. During that academic year I was grateful to Columbia University who gave me the title of Visiting Research Professor and gave me the freedom to work in their many well-stocked libraries.

Since then I have been engaged in the process of writing up my research, which has become a large project examining Hotelling's role in the history of economics. This volume is the first fruit of those labours. It comprises a lengthy introductory chapter which seeks first to describe Hotelling's life, and then proceeds to identify the nature and importance of his seminal contributions to economics. His published articles in economics (excepting two obituaries, of Schultz and Roos, an entry in *Encyclopedia Americana* and all his book reviews) are reprinted here, having corrected only original typographic errors; all page references have also been changed to the pagination of the reprints.

His work covers diverse fields of economics, but is unified by its use of sophisticated mathematical techniques. This collection will, hopefully, be of interest to all economists, but especially to those whose research lies in the history of our subject. Additionally, many economists today find inspiration for their work in the path-breaking papers Hotelling published some fifty years ago, and it is to be hoped that this collection will itself provide further inspiration for research in the general field of mathematical economics—the field in which Hotelling was certainly one of the most successful twentieth-century pioneers. A reading of his papers will not prove relaxing, necessarily, but it will prove most rewarding.

Acknowledgements

First and foremost I would like to thank both Mrs. Harold Hotelling and Professor Harold Hotelling, Jnr. for their most generous and kind assistance and encouragement in this research. Additionally, I would like to thank the staff of the Butler Library and particularly the Librarian of Rare Books and Manuscripts, Dr. K. Lohf, for all the assistance given; I also acknowledge the help given by the staff at the Archive of The Food Research Institute, Stanford.

I am indebted to Professors Kenneth Arrow, Mark Blaug, Denis O'Brien and Ingram Olkin, and to Dr. J. Lynne Evans for their encouragement and for their comments upon my work. Last, but by no means least, my thanks go to Angela Darnell for her invaluable work as an unpaid research assistant on a preliminary visit to Columbia in April 1986.

I acknowledge, with most grateful thanks, the permission given by Mrs. Harold Hotelling and by Dr. K. Lohf to quote from the papers in the Hotelling Collection at Columbia University, and I also acknowledge the co-operation of the publishers of Hotelling's articles who have granted permission to reprint copyright material originally published by them. All work reprinted is without alteration from the original, with the exception of the correction of typographic errors and the imposition of internally consistent pagination.

Generous financial assistance from the Nuffield Foundation (Grant No. SOC/181(1412), the British Academy, the Economic and Social Research Council (Grant No. B 0023 2184) and the University of Durham is gratefully acknowledged.

Contents

The Life and Economic Thought of Harold Hotelling

I. INTRODUCTION

Harold Hotelling was one of the most important of the twentieth-century pioneers of mathematical economics and mathematical statistics.[1] In both fields he is recognised for his powerful theoretical contributions[2] and he was a most effective and caring teacher. His influence, both as an economist and as a statistician, is felt not only through his publications (a large number of which are seminal[3]), but also through his students, amongst whom one can count many of the leading economists and statisticians of the next generation.[4]

Hotelling's career spans one of the most creative and formative periods in the development of both mathematical economics and mathematical statistics, and few figures have displayed a comparable originality; fewer still have publication records which bear comparison. Most importantly, his published papers are today seen as the starting point of much contemporary research. Indeed, his name is familiar to a remarkably wide range of professionals,

---

[1] There are, surprisingly, few secondary sources on Hotelling; additional details on his life and career may be found in Madow (1960), Samuelson (1960), Smith (1978), Pfouts and Leadbetter (1979), the *American Economic Review* (1974), Arrow (1987) and Darnell (1988).

[2] His numerous papers in mathematical statistics did have immediate impact; however, during his own lifetime, many of his contributions to economics were not properly recognised (for reasons discussed in this introductory chapter). His publications (in all fields) are listed in a complete bibliography at the end of this work. This current volume is concerned with his writings in economics, and of those, only his work in spatial economics (1929) and in welfare economics (1938, 1939 and 1939a) may be said to have had contemporary impact.

[3] His publications in economics are the real concern of this work; see Hotelling (1925, 1929, 1931, 1932, 1933, 1935, 1936, 1938, 1939, 1939a and 1943), all of which are reprinted immediately after this introductory chapter. However, in the field of mathematical statistics see, especially, the seminal contributions in Hotelling (1931a, 1933a, 1936a and 1940), and for an appreciation of Hotelling the statistician, see Neyman (1960) in Olkin, Ghuye, Hoeffding, Madow and Mann (1960).

[4] The list is far too large to catalogue. Nevertheless, suffice to note that while at Columbia, he taught Friedman statistics, supervised Arrow's PhD, and worked with Stigler in the Statistical Research Group.

a range which includes, naturally, economists and statisticians but which also includes educationalists and psychologists. He is, however, best known as a mathematical economist and mathematical statistician: in economics there is Hotelling's Law,[5] Hotelling's Lemma[6] and Hotelling's Rule,[7] and in statistics there is Hotelling's T^2 statistic.[8]

II. Biographia

Hotelling's biographical details are not widely available,[9] yet they provide a significant insight into his motivations as an economist and into his explicit dependence upon mathematical methods.

Harold Hotelling was born on 29 September 1895 in Fulda, Minnesota, of ancestors long American but originally of English and Dutch extraction. When he was nine his family moved to Seattle in order that the five children, of whom Harold was the oldest, might enjoy a better education and have access to better opportunities than those available in Fulda. He attended high school in Seattle and, on leaving, began studying journalism at the University of Washington, financing his studies working for small newspapers; he further supplemented his income by carrying out small re-wiring jobs having studied the rudimentary principles of electricity through his own private reading (rather than through any formal tuition). His studies were interrupted by the First World War and he notes n a biographical sketch[10] that, on being called to war service:

> Having studied mathematics, science and classics at school and college, was considered by [the] army authorities competent to care for mules. The result was [that] a temperamental mule temporarily broke his leg and thereby saved his life, as the rest of the division was sent to France and [was] wiped out.

Thus Hotelling was, fortuitously, never engaged in active service and he resumed his course in journalism having been discharged from the army on 4 February 1919; however, Hotelling found that the Department of Journalism

---

[5] The Law refers to the observation that spatial competition may lead to a clustering of competitors. See Hotelling (1929).

[6] The Lemma refers to Hotelling's work on integrability and duality in the context of the theory of the producer, and refers to the fact that the first partial derivative of a cost function with respect to the price of factor i is the demand function for that factor. See Hotelling (1932).

[7] The Rule refers to Hotelling's work on exhaustible resources, and states that the optimal production schedule of an irreplaceable resource is such that it generates a rate of increase of its price which is identical to society's discount rate (thus if society's discount rate is r, then the price of an exhaustible resource should also rise at rate r). See Hotelling (1931).

[8] The T^2 statistic is used to make simultaneous tests of the equality of several characteristics of many variables. It is a generalisation of Student's t statistic. See Hotelling (1931a).

[9] See the references given in footnote 1.

[10] This sketch was written cryptically and in the third person as a press notice for an invited lecture Hotelling gave in 1932 for the Chemical Society of New York. Hotelling lectured on the title of "The economics of obsolescence".

had been severely disrupted by the war and he began to study some economics. In a letter to the University of Washington, dated 14 December 1962, he wrote:[11]

> I have a first degree in journalism from Washington but at the time the School of Journalism was badly disorganised because of the First World War and they let me substitute so much economics for journalism that the degree might just have well been in it. Anyhow, the journalism degree got me a job on the *Washington Standard* in Olympia. Its economic component, plus later studies of mathematics at Seattle, Chicago and Princeton gave enough color to the idea that I was an economist to give me a job as Professor of Economics at Columbia University. I actually did teach economics there, but it was economics so mathematical that no member of the distinguished economics faculty there could understand it.

In the light of consultation with university records. Hotelling's own report makes puzzling reading: he earned a total of only nine credits in economics from just three courses. One was in "Railway and Marine Rates" and another in "Marine Insurance"; the description of the third is, unfortunately, not available. This represents only a very small part of his degree, being less than three-quarters of his final year studies.

He was awarded his BA in journalism in 1919 and joined the staff of a local newspaper. Hotelling records in other autobiographical notes[12] that he was stimulated as a very young man by "[the family's] Methodist interest in social justice and race equality, and the family tradition of active concern with public problems.",[13] and it was with these sentiments that he became a journalist. He soon became disillusioned:

> Journalism had seemed to offer both a means of livelihood and an opportunity to stimulate proper action on public matters. Later I concluded that it had been overrated in both respects.[14]

Much of his disappointment stemmed from his inability to write sufficiently quickly and many of his pieces were not published due to their consequent lack of topicality. With the encouragement of one of the teachers at Washington, Eric Temple Bell, the great mathematician and historian of mathematics,[15] Hotelling returned to the university on 1 January 1920 to study mathematics

---

[11] This letter was written by Hotelling in answer to an enquiry from the University of Washington who were, at the time, seeking information on their celebrated alumni.

[12] These autobiographical notes come from "Western Hotelling and Allied Families on Epic of Migration", an unpublished, undated, typescript Hotelling wrote which traces the history of his family and background. It is better described as a family history than as an autobiography of Harold Hotelling.

[13] Hotelling was never an active church-goer; he was, however, a teetotaller and always looked unfavourably upon the consumption of alcohol.

[14] Also quoted from "Western Hotelling ...".

[15] Bell is particularly well known for his work on the history of mathematics, notably his *Men of Mathematics* (1937).

at master's level, financing his studies by what he described as a badly paid job of teaching trigonometry and analytical geometry. At this stage in his career, Hotelling was learning (and teaching) mathematics for the first time since leaving high school. He attended a mathematics summer school at Chicago in 1920[16] and obtained his master's degree from Washington at the end of 1921; ironically, as later events were to prove, he applied, in the spring of 1920, for a fellowship in economics at Columbia but was rejected. He was, however, successful in his application for a fellowship in mathematics at Princeton.

What Hotelling actually wanted to do was to study mathematical economics and statistics, and the purpose of his mathematical studies at Washington was simply to lay the foundations of this ultimate goal. This was all a part of a long-standing strategy. He wrote that:[17]

> The combination of science and political economy led to the thought of applying the methods proven so useful in the exact sciences to discover new truth in economics and political science. Proficiency in these methods required in the first place mathematics.... Ideas of flow and diffusion in mathematical physics looked as if they might have applications to human migration and to the flow of commodities.[18]

Having gained the MA in mathematics and thus achieved proficiency in what he felt was the mathematics necessary to study economics, Hotelling was then looking for an opportunity to develop his understanding of economics; however, in going to Princeton he was initially frustrated in this aim:

> In going to Princeton I had intended to study mathematical economics and statistics; actually I found there was no one there who knew anything about either subject. I therefore studied the topology, differential geometry, analysis, mathematical physics and astrophysics that Princeton then offered, and all of these have to some extent contributed to my later work.

He pursued his doctorate studies at Princeton, concentrating on Analysis Situs (better known today as topology) under the direction of Oswald Veblen, the nephew of Thorstein Veblen, and was awarded his degree in June 1924. Although Hotelling was frustrated in his ambition of receiving formal tuition in mathematical economics, he was able to pursue his objective through active membership of the American Mathematical Society (AMS).

The AMS had been formed in the late nineteenth century to provide a

---

[16] During the summer of 1920, one of the Chicago schools was given by G. A. Bliss, one of the mathematicians responsible for developing and popularising the calculus of variations technique. I have, however, failed to confirm that these classes were attended by Hotelling, although he does make extensive references to Bliss' work in the unpublished manuscripts. I think it most likely that Hotelling did attend Bliss' summer lectures.

[17] This, and the following quotation, are taken from Hotelling's own "Western Hotelling...".

[18] The concepts of flow and diffusion were central to his master's degree thesis which utilised them in a model of migration patterns.

national and regional forum to facilitate the dissemination of both theoretical and applied mathematical research, and had amongst its members many of the foremost mathematical economists of the 1920s. Indeed, the AMS was more of a forum for mathematical economics at this time than was the American Economic Association (AEA).[19] Hotelling first attended the AMS in April 1923, when he was presented to the society by the then president and his doctorate supervisor, Veblen. The meeting was held at Columbia University, and Hotelling was elected to membership; he attended the next two meetings of the Eastern Section at Columbia, in October and December 1923, and became a regular attender of regional and national meetings for the next decade.

On completing his doctorate, Hotelling was appointed as a Junior Associate at the Food Research Institute (FRI), Stanford; he acted as a mathematical consultant, offering advice to any of the institute's staff who sought it.[20] Prior to Hotelling's appointment, the institute had called upon members of the Mathematics Department for consultation on matters outside the expertise of the institute's staff; however, by 1924 it became apparent that there was sufficient work to warrant a full-time mathematician who also had statistical skills. Hotelling was an impressive candidate for the post, both as a technically skilled mathematician and as someone with a genuine concern for economic issues. Veblen recommended him highly and he was appointed to the post for one year in the first instance.

The primary task for which Hotelling was responsible was to aid in the work of estimating crop yields and food requirements, especially those for wheat. This task required statistical expertise, in which Hotelling had had no formal tuition, and, on the suggestion of Carl Alsberg, one of the institute's directors, Hotelling began reading the work of R. A. Fisher. This had a profound effect upon Hotelling's future work in mathematical statistics, much of which may be seen to have a marked Fisherian flavour. Indeed, Hotelling and Fisher entered into correspondence on statistical matters, notably concerning issues of correlation and the probabilistic foundations of statistics, and Hotelling secured a leave of absence from Stanford in the fall of 1929 which

---

[19] In fact, the AEA at this time was a somewhat hostile environment for any economist utilising mathematical methods, and few "mainstream" economists would have been able either to understand or appreciate the results achieved by the newly emerging mathematical economists. An examination of issues of the *American Economic Review* from this period illustrates the marked lack of published mathematical papers. In their study of the use of mathematics in published economics papers, Grubel and Boland (1986) reported that the marked acceleration of the use of mathematics occurred around the early 1960s. They observed that "In 1951 only 2.2 per cent of all pages [of the *American Economic Review*] contained at least one equation. By 1978 this proportion had risen to 44 per cent." (p. 425).

[20] The FRI had been established in 1921 as a result of co-operation between Stanford University and the Carnegie Corporation with the general brief to build up and make available an accurate, and increasingly comprehensive, body of knowledge about the economic determinants of welfare, especially as they pertained to agriculture and activities related to it.

he spent all the Rothamsted Experimental Station working with Fisher.[21]

Hotelling worked at Stanford for some seven years, although he gradually moved from the FRI into the Mathematics Department. In February 1926 he was appointed to a joint post in the Mathematics Department and the FRI and stayed in this position for one year. He was involved in very little teaching and divided his time between the two units; he also acted as a mathematical consultant to members of the Economics Department.[22]

In 1927 Hotelling accepted a full-time assistant professorship in mathematics and there he remained, teaching a unique combination of courses, including topology, statistics and differential geometry, until 1931 when he moved to Columbia. Hotelling was both very happy and very successful at Stanford[23] and during his seven years there he was approached by other universities who sought to attract him to their Mathematics Departments.[24]

---

[21] The two worked very closely together, and they began writing a basic statistics textbook together. This project was abandoned in the early 1930s, due not only to the difficulties of pursuing joint work by correspondence but also to the fact that their thinking diverged on some fundamental issues in statistics. This project was then pursued over the next twenty-five years as a singly authored work by Hotelling; his students were aware of Hotelling's involvement in writing such a book, and, not surprisingly, his introductory lecture course bore close correspondence with the early drafts. However, he never devoted himself to the book whole-heartedly, and by the time he had finished the "last chapter", he felt that the earlier chapters then needed revision in order to bring them up-to-date. This process had no end, and although he was asked by a number of publishers to supply them with the manuscript, it was never in a sufficiently final form to satisfy Hotelling. The friendship and professional association between the two was long-standing, and Fisher as one of the many distinguished visitors to the Statistics Institute at Chapel Hill when Hotelling took up his appointment there.

[22] The Economics Department was most enthusiastic about this arrangement for it gave its members the opportunity to call upon Hotelling's mathematical expertise. Indeed, the Economics Department paid $500 per annum towards Hotelling's salary, in return for which they had formal access to Hotelling as an *ad hoc* consultant. The then Head of Economics (John Canning) recognised that there were:

> statistical problems in the field of economics that may be expected to be solved only by a first rate mathematician who is also an economist. We know also that men possessing these joint qualifications are rare. (Canning to Green, then Head of the Mathematics Department, Stanford, 3 February 1926.)

[23] He published many papers during this period, some from his PhD (see Hotelling (1925c and 1926) for example), some in mathematical statistics (see Hotelling (1925d, 1926a and 1930) for example) and two other papers of great importance. One, Hotelling (1927), was concerned with differential equations subject to error and, though written in the context of population problems, is a significant contribution to the general theory of stochastic processes. The other, Hotelling and Working (1929), dealt with regression problems in the context of population data and showed that the standard error of an estimate depends upon the value of the independent variable.

[24] In particular, he was clearly tempted by offers from Michigan and Brown. The prospect of moving to the East was most attractive, but neither offer was sufficiently tempting. For example, during 1927 he was in correspondence with James Glover, the then Chairman of the Mathematics Department at Ann Arbor, Michigan; Hotelling wrote on 27 October that he had "a considerable desire to be nearer ... to the center of mathematical activities for the sake of easier communication and readier stimulus to research", but Glover's budget could not match Hotelling's demand of a minimum salary of $5,000 per annum and a full professorship.

It was not until 1931 that Hotelling was made an offer he felt able to accept, and in that year he moved to Columbia University as Professor of Mathematical Economics in succession to Henry Ludwell Moore who had been forced, by ill-health, to retire at his own request from 1 April 1929. Just after the move to Columbia, Hotelling's wife fell ill and she died in October 1932.[25] The immeasurable loss of his wife's sudden death fell at the beginning of what was to be Hotelling's most productive period as a publishing economist; the task of looking after his two small children was, however, eased by his sister-in-law who, for the next few years, cared for them with her own children who were of similar ages.[26]

Hotelling taught both mathematical statistics and mathematical economics at Columbia[27] acted as a consultant to various outside agencies[28] and, in 1943, he also became Operations Director of the newly created Statistical Research Group at Columbia.[29] Hotelling was closely associated with all its work, spending just over three years in a part-time capacity as Principal Investigator. Additionally, while at Columbia, Hotelling had continually and consistently attempted to establish a statistics institute, and had also attempted to have statistics seen as a discipline in its own right, as opposed to a service subject to other disciplines.[30] In both attempts he was frustrated, but after the war he was approached by Gertrude Cox, the director of the newly

[25] In 1920 he had married Floy Tracy. Two children were born of this marriage, the first was born in 1923 and was named Eric Bell, in honour of the mathematician "who had foreseen the direction my career should take and over a series of years had supplied the guidance and inspiration to offset many diverting cross-currents." (Quoted from "Western Hotelling ...".)

[26] In June 1934 Hotelling remarried. He had first met his second wife, Suzanna Porter Edmondson, when she was a student in his statistics class at Columbia in 1932. She was born in Montgomery, Alabama, in 1909 and was educated in various colleges, where she studied modern romantic languages. She won several scholarships which enabled her to visit France, but in the early 1930s she decide upon a career change and went to Columbia to study statistics. On marrying Hotelling, she abandoned her academic studies and devoted herself tirelessly to supporting him in his established career. Six children were born of this marriage. She acted as hostess to the many distinguished guests who stayed with the Hotellings and, every second Sunday of the month they held an "open house" for Hotelling's students, colleagues, visitors and families. These events were renowned and were known as "Hotelling Teas", with pun intended after the T^2 statistic.

[27] However, Columbia in the 1930s was a somewhat hostile environment for such courses; "The predominant interests of the Columbia Department of Economics were actively anti-theoretical, to the point where no systematic course in neoclassical price theory was even offered, let alone prescribed for the general student." (Arrow (1988), p. 671.)

[28] Of his many consultancies, mention should be made of the New York Telephone Company, the National Recovery Administration, the Division of Tax Research of the Treasury Department, the Wartime Office of Scientific Research and the Bureau of the Budget.

[29] The Statistical Research Group had been formed to address various statistical problems which arose as a consequence of the war effort and "was composed of what surely must be the most extraordinary group of statisticians ever organised, taking into account both number and quality." (Wallis, 1980). See also Anscombe (1980) and Kruskal (1980). Among the problems examined was a statistical approach to the quality control of bomb sight manufacture. A number of publications resulted; see, especially, Hotelling (1947).

[30] These views were articulated in his publications; see, especially, Hotelling (1940).

founded Statistical Institute, North Carolina. She was looking for a co-director and wanted Hotelling to take up this position. At this time, Hotelling was one of the leading mathematical statisticians of the day and, even though he was not actively seeking a move from Columbia, he was attracted by the opportunity of becoming co-director of precisely the sort of unit he had unsuccessfully tried to form at Columbia.[31] He moved to North Carolina in 1946 to take up this new position and remained there for the rest of his life.

He published very little while at Chapel Hill but was, nonetheless, active in research, in economics to some extent, but more so in mathematical statistics. Primarily, however, he devoted himself to the responsibility of establishing the institute as the premier centre of excellence in statistics, attracting the then leading statisticians to its staff and attracting a set of graduate students who were to become the leaders of the next generation of statisticians. He taught extensively and had the titles of Professor of Mathematical Statistics and later became Kenan Professor of Economics. He retired first in 1960 (when he was 65) but carried on teaching until he was 70 when he finally retired. Hotelling died on 26 December 1973 of a stroke after an illness of eighteen months.

Hotelling's work in economics covers the period from the early 1920s through to the 1960s, but his publications cover the much shorter period from 1925 to 1943; of the research in economics he completed after his move to North Carolina, none was published. Indeed, Hotelling was never a prolific publisher of economics: his total output over the 18 years from 1925 numbers eleven articles.[32] Hotelling never published a book. Although he was a mathematical statistician of the first order, making several contributions of fundamental importance, this current volume seeks to highlight the contribution made by Hotelling as an economist.

He received numerous honours during his lifetime, but one of the major tributes to his stature as an economist came in 1965 when the American Economic Association, having created the honour of Distinguished Fellow, elected Hotelling to be the first recipient of this most prestigious award.[33]

---

[31] Hotelling had originally been asked by Cox what offer would tempt him to move from Columbia, and he replied that the salary of the country's top football coach would tempt him. This was relayed to the President of North Carolina who agreed to this. Hotelling's "demand" having been met, he felt bound to accept the offer; Columbia, on receiving his resignation offered then to enter into negotiations, but Hotelling's sense of honour led him to refuse this. Thus he left to become the Co-Director of the Statistical Institute. It should also be added that his lack of success in creating a statistical institute at Columbia was in part the consequence of his lack of political acumen within the university. He was not accomplished at dealing with the university's administration and failed to acquire the resources necessary to set up the new venture. Within two years of his leaving, however, Columbia established a Statistical Institute. The groundwork had been laid by Hotelling, and Columbia did not wish to be seen as "behind the times".

[32] Excluding two papers which were obituaries, of Schultz and Roos, (Hotelling (1939b and 1958)) one entry in *Encyclopedia Americana* (Hotelling (1955)) and all book reviews.

[33] This was a joint election together with Edward E. Chamberlin. It is most interesting to note that Hotelling had not, until this point in time, chosen to be a member of the American Economic

Hotelling's contributions to economics range over a wide area and are unified by their use of mathematical tools of analysis which enabled him to derive results which had, until then, been inaccessible to economists who were using less formal techniques. Of his published papers, the best known are perhaps those of 1929, on "Stability in competition"; 1931, on "The economics of exhaustible resources"; and 1938, on "The general welfare...". Each of these papers has formed the basis of much contemporary research, but it should not be inferred from this that the remaining papers lack merit; indeed, the three papers on the nature of supply and demand functions are extremely important and in these may be found the foundations of duality theory and some notable contributions to the theory of integrability. His published papers may conveniently be grouped into those concerning depreciation (1925, 1931),[34] competition (1929), supply and demand functions (1932, 1933 and 1935) and welfare economics (1936, 1938, 1939, 1939a and 1943). This categorisation will be used below.

III. Depreciation

At one of the first AMS meetings Hotelling attended, in December 1923, a Dr. James S. Taylor presented a paper on "A statistical theory of depreciation based on unit cost" in which he examined the questions of how to determine the useful life of a machine and how to distribute the depreciation charges optimally over time; in so doing, Taylor identified the interdependence between the costs of production and the distribution of depreciation, since the former necessarily includes, as a component, the latter. Taylor also noted that the useful life of a machine is in part determined by the ruling interest rate.[35]

Taylor used a discrete time framework and Hotelling chose to develop this analysis within a continuous time frame. By so doing he was able to extend Taylor's work and correct what he perceived as some of its shortcomings. Hotelling published this in the *Journal of the American Statistical Association*

Association. This may be explained in at least two ways. First, Hotelling's methods were mathematical, and as noted above, the AEA contained few mathematical economists at the time he was making his contributions; second, Hotelling felt that the mathematical approach was **the** way forward to a "scientific economics", and the small number of economists prepared to use mathematics was viewed by him as indicative of a lack of contemporary success of this approach (especially in the 1930s). He felt that his pioneering efforts, and those of such economists as Evans and Roos, had been ignored and this led to his being somewhat disillusioned with the profession at large. In fact, the citation from the AEA for the award made special mention of his presidential address to the Econometric Society in 1938 concerning welfare economics (see Section VI), and this paper was the only one of his contributions which enjoyed an immediate acclaim and which sparked a contemporary debate in the literature.

[34] There is, additionally, a small piece written by Hotelling and published in 1940; this was a consequence of his having refereed an article by Preinreich (1940) for *Econometrica*, and some of the points he raised then became a "question and answer" appendix to Preinreich's paper. See Hotelling (1940a).

[35] See Taylor (1923) and (1925).

in 1925, in a paper which was described by the (anonymous) writer of his obituary in the *American Economic Review* as "a turning point in capital theory proper and in the reorientation of accounting towards more economically meaningful magnitudes" (p. 1102). However, this publication represents only a very small part of Hotelling's research accomplishments at that time. The published work was a development of a paper he had presented to the AMS in December 1924, entitled "Economic problems involving maxima of functionals" (1925a). This was of fundamental importance to much of his later work.

From the original notes, it is clear that Hotelling first set out the mathematical foundation of his then current work on depreciation of a replaceable asset, pointing out that the problem could be reduced to a problem in ordinary differential calculus; he went on to observe that many problems in economics could be reduced to problems involving the maximisation of functionals and, therefore, to problems in the calculus of variations.[36] That the problem of depreciation could be reduced to a simple problem involving the differential calculus is a consequence of assuming that at all points in time the machine is used to full capacity. A specific example, quoted by Hotelling, where this assumption is unwarranted is when the machine in question is not replaceable, and analysis of such a situation will lead to conclusions regarding the optimal **path** of usage; additionally, the analysis of the optimal use of a non-replaceable machine whose total capacity for production is known will generate, as its solution, statements of both the optimal life of the machine (expressed in time, as opposed to total production) and the optimal output stream at each point in time. By substituting the word asset for machine, a general framework in which the economics of exhaustible resources may be analysed is obtained, and the resource in question may be any physical asset such as a machine or a coal mine. It was this more general problem that Hotelling addressed in the paper presented to the AMS in December 1924.

Questions which involve the analysis of optimal strategies over time involve the mathematical technique of dynamic optimisation, and may, typically, be approached using the calculus of variations. This technique was then little used by economists (and is still relatively little used by economists); however, without this mathematical tool[37] the full intricacies of the problem in question are inaccessible. Interestingly, given that Hotelling's audience comprised mathematicians, he began the paper[38] by providing a very detailed derivation of the solution to a general calculus of variations problem,[39] and this can

---

[36] Problems involving functionals may be characterised as those which involve functions whose own domains are themselves functions.

[37] Or some equivalent technique such as optimal control theory and the use of Hamiltonians.

[38] Various drafts of this paper are to be found in the Hotelling Collection at Columbia. An interesting observation on the drafts, and upon the finally published version in the *Journal of Political Economy*, is that there is no reference to the earlier work of Gray (1914) on exhaustible resources; Hotelling was, no doubt, simply ignorant of its existence.

[39] This derivation included both the first-order Euler–Lagrange condition and also examined second-order conditions and the Weierstrass condition.

only be taken as an indication that he felt the technique was insufficiently familiar even to such an audience.

The paper proceeded to derive the solutions to the problem of mining economics first under what he called "competitive" conditions, and then under monopoly conditions. The final section of the paper suggested the very general applicability of the calculus of variations in economics: "all hedonistic and eudaemonistic theories reduce the problem of right conduct to that of maximising functionals" he wrote. The abstract of this paper (see Hotelling 1925a) is remarkable for the achievements claimed for the work:

> The very puzzling problem of mining economics reduces by an immediate integration of the Euler equations to a system of first-order differential equations. This solves the problem of depreciation of irreplaceable assets, both under competition and under monopoly.

Hotelling used three assumptions in deriving the optimal output path: first that production was costless, second that the resource would be exhausted and, finally, that at the terminal time which was to be determined, the output rate would be zero. Each is a simplifying assumption, and each can lead to a loss of generality; much recent research has been directed towards their relaxation. Nevertheless, this should not be seen as detracting from the enormous achievements of the paper; indeed, the results derived regarding the optimal output schedule for a monopolistic mine owner are identical to those reported some seven years later in his famous 1931 *Journal of Political Economy* paper. The two additional major issues examined in the 1931 publication, namely the optimal rate of production for a price-taking owner, and the optimal rate from society's point of view, were not considered at this early stage of Hotelling's work on the subject.[40]

The analysis of the 1924 paper concerning monopolistic production was, within the assumptions noted, complete for the case of costless production, and the general results were applied to a number of worked examples which utilised various specific linear and non-linear demand schedules. The case of "competition" did not allow any such definitive results, and for very specific reasons which are described below.

As has been stated, Hotelling did not enjoy much formal training in economics; records indicate that his undergraduate teaching was confined to three final year courses, and it is of interest to enquire of the source of his economics. Some clues may be derived from his use of the word "competition": in 1924 his understanding of "competition" was within the French tradition and had no connection with the later concept of "perfect competition". Furthermore, in all his work, Hotelling eschewed the Marshallian analytical methods of partial equilibrium, and chose to consider problems within a more general equilibrium framework. This may be traced to Hotelling's professional

---

[40] However, in another paper of 1925, devoted exclusively to the problems of mining economics, the issues of optimal extraction from society's point of view and the extraction rate adopted by a price-taking industry were both considered. (See Hotelling (1925b).)

association with Griffith C. Evans, a mathematician at the Rice Institute in the 1920s who later joined the Mathematics Department at Berkeley. Evans published several articles in economics in the 1920s, published a pioneering text in mathematical economics in 1930[41] and was a founder member (as was Hotelling himself) of the Econometric Society in the early 1930s. Hotelling's contact with him was through the AMS and through his publications. Evans was himself much influenced by Cournot, and many of his papers presented mathematical extensions of Cournot's work.[42] To such writers as Hotelling and Evans, "competition" meant a system of price-setting producers within an industry comprising so few firms as to require an industry analysis incorporating all the inter-dependencies. Thus, Hotelling in 1924 considered the case of "competition" by which he meant an industry comprising n firms, each of which makes decisions about its own production schedule on the assumption that it faces a downward sloping demand curve and on the assumption that its output decision will not induce a quantity reaction from the other $n - 1$ firms. This is wholly within the Cournot tradition. In some unpublished notes of 1925, he observed that:

> The selection of production and price policies, when not arrived at through collusion is made by a trial-and-error process. One owner varies his rate of output q so as to increase his own J [the net present value of all future profits, including royalties]; this affects all other J's, leading the other owners to vary their q's. Each is affected by every variation of rate of output and consequent change of price. Without entering into the interesting question of economic dynamics involved, we here suppose merely that the process has converged to the final adoption by each owner of a settled policy, and enquire what the equilibrium policies are.

In other notes, also of 1925, Hotelling examined the case of duopolistic owners of exhaustible resources and, not surprisingly, was unable to determine a solution; as he observed, once one mine is exhausted, the remaining mine has a monopoly, and it need not be the case that the larger mine will last the longer.

That Hotelling, little trained in economics, had been able to derive so many results of importance using a relatively unfamiliar technique is remarkable; he brought much of this research together and extended it in a paper he delivered to the San Fransisco section of the AMS, meeting at Stanford in

---

[41] Evans' text, *Mathematical Introduction to Economics* (1930), is a quite remarkable achievement, but is not widely known. In the UK it received two hostile reviews (see Allen (1931) and Bowley (1932)) and was little used as a consequence. Hotelling reviewed it favourably (see Hotelling (1931b and 1931c) as did Wright (1931). Hotelling remarked that "This book helps to lay a groundwork upon which future contributions to political economy of first rate importance may be expected to be based." (1931b, p. 109); unfortunately, the impact was far from immediate due to the largely non-mathematical training and inclination of the majority of economists on both sides of the Atlantic.

[42] See, for example, Evans (1922, 1924, 1925, 1925a and 1930). For more detail of the work of Evans, his pupils (especially Roos), and the AMS, see Darnell (1982).

April 1925. This paper, entitled "Theory of mine economics", represented an extension of his earlier work by introducing costs to the anslysis via the device of a "price net of extraction, transportation and marketing expenses" and the notion of a "royalty" was introduced. However, Hotelling did not introduce a "stock effect"; thus costs were dependent only upon current output, and were independent of accumulated extraction.[43] In much later work, dated 1934 and never published, this assumption is relaxed, leading to the result that at the terminal time, the mine need not be exhausted.[44] However, the paper he presented to the AMS in April 1925 is also worthy of note for its extension of the analysis to a consideration of **society's optimal** rate of extraction; in the final section of the presentation, he observed that:

> Though natural resources may in some cases be exploited too rapidly under competition [in the sense of price-setting competition among a few suppliers], the analysis shows that in other cases the exploitation is slower than the public interest requires.[45]

The latter is a reference to the fact that a monopolist will, under particular assumptions, produce at a rate slower than society's optimum. An industry comprising price-taking firms will produce at society's optimum rate if the individual firms each use an identical discount rate which itself is identical to society's discount rate.[46]

This work was brought together by Hotelling into his famous 1931 *Journal of Political Economy* article; given that many of the paper's results had been worked out some years prior to their publication, it is of interest, and of some importance, to note that the paper had, according to his own later accounts, been originally turned down by the *Economic Journal* on the grounds that its mathematics were too difficult.[47] The rejection on these grounds is, perhaps, a little surprising since the then editors (Keynes and MacGregor) had published Ramsey's two papers on optimal taxation (1927) and on growth (1928), the latter of which used precisely the same mathematical technique (the calculus of variations) as Hotelling's work.

The importance of Hotelling's paper is great; indeed, all subsequent work

---

[43] There is a stock effect in Hotelling (1931), but there he assumes that, at the terminal time (possibly infinite), the resources will be exhausted. (See p. 153 especially on this point.)

[44] This result, which introduces the notion of the "economic life" of a mine, is dependent upon the size of the market and the nature of the cost function. In the absence of any stock effect, that is, when the costs are independent of the amount of resource extracted to date, the resource will be exhausted at terminal time (which may, however, not be finite).

[45] See the abstract of this paper, Hotelling (1925b).

[46] For formal derivations of these propositions, see Hotelling (1931, especially pp. 151–2) and see, for example, Conrad and Clark (1987, especially pp. 117–23).

[47] This was related to me by Kenneth Arrow and it is also reported in the 1974 *American Economic Review* obituary of Hotelling; Arrow believes that he was one of the few to have known of this rejection, but does not remember having written the obituary, although he thinks it "more than likely" that he was its author.

on the problem has been essentially based upon this work, and the literature recognises it as the seminal paper on the subject.[48] Hotelling himself was particularly pleased with this paper, but perhaps more because of its use of mathematics than for its economic content: in a letter of 28 August 1965 he described the paper as "one of the major triumphs of my career as I found effective use in it for the calculus of variations".[49]

IV. SPATIAL COMPETITION

Hotelling's published work on spatial competition is confined to his one paper (1929), and there are no records of his early analyses, except for a fragment entitled the "grocery store problem", dated 21 November 1924. This set out the simple analytics of what was to become the published paper of 1929, but in a most cryptic fashion. Hotelling actually presented an early version of his published paper to the AMS in New York on 6 April 1928, and this was revised for submission to the *Economic Journal*. On submission, Keynes, as editor, accepted the paper subject to some minor revisions which consisted of Hotelling responding to Keynes' request that reference be made to Sraffa's work on market segmentation.[50] Hotelling's failure to refer to this work in the submission may well represent his ignorance of the economics literature; in the original version, he had made reference to Edgeworth, Cournot, Amoroso, Evans and Roos, a list which provides a fair indication of the extent of his reading in the economics. It is to be noted that in all his published work, Hotelling was sparing in his references to other economists; this is a consequence both of his knowledge of the literature and of the relevance of the literature. Having had little formal training in economics, his reading was necessarily limited but, more importantly, Hotelling's work had few precursors and hence there was little opportunity or need to make extensive references. As a mathematical economist, Hotelling was working at the very limits of the subject and spent much of his time "translating" the work of earlier economists into the language of mathematics; in the process he was able to identify the lack of generality of some arguments and was also able to identify areas for further research.

The paper of 1929 represents a seminal contribution to the theory of spatial competition and the theory of locational equilibrium. The paper was essentially an exercise in game theory, in which, at the first stage, the two players choose a location in a linear market, and in the second they each choose a price. Hotelling described the equilibrium of such a game, but unfortunately his conclusion is in error since his equilibrium is only local.[51] Nevertheless, his

---

[48] See, for example, the assessment in Devarajan and Fisher (1981) and see, also, Peterson and Fisher (1977).

[49] Hotelling to Burt, 28 August 1965. Oscar Burt was at the Department of Agricultural Economics, Missouri.

[50] Sraffa's work refers to segmentation of markets in such a way that within each "sub-market" one seller can adopt a quasi-monopolistic position. See Sraffa (1926).

[51] This was pointed out some fifty years later by d'Aspremont, Gabszewicz and Thisse (1979).

paper is a true landmark in the area and is regarded as the seminal contribution to the subject. Although Hotelling's paper stood alone in the field for many years, all subsequent forays have found inspiration in his analysis, and it has been the source of a large literature.[52] Hotelling returned to this topic in 1934 when Lerner and Singer submitted a paper to the *Journal of Political Economy* which extended the Hotelling model.[53] He was invited, on acceptance of their paper, to submit a response. This was partially written, though never published, and comprises a typed manuscript of six pages plus ten pages of handwritten notes.

This later manuscript extends the analysis of the 1929 paper by replacing the original assumption that demand is perfectly inelastic with the assumption that demand is simply some monotonic decreasing function of the price of the good. The main conclusions of the earlier paper remain unchanged by this extension, namely:

> an uneconomic tendency for competitors to fix their locations, and likewise the qualities of their wares, in a manner quite different from that which would bring about a maximum of social efficiency.

Hotelling used the concept of "location" in a figurative sense, so that it could be interpreted as any dimension of the product other than price; for example, "location" could be literally interpreted as indicating a geographical quality of the goods, or it could be interpreted as some other dimension of quality. In the original article, he noted that the analysis could have a political interpretation:

> The competition for votes between the Republican and Democratic parties does not lead to a clear drawing of the issues, an adoption of two strongly contrasting positions between which the voter may choose. Instead, each party strives to make its platform as much like the other's as possible. (p. 54).

And the closing sentence of the article reads: "Methodist and Presbyterian churches are too much alike; cider is too homogeneous." (p. 57). This broad conclusion was reiterated in the later manuscript of 1934: "[the result of this analysis] might mean physical clustering in space, a wasteful tendency to imitation in various qualities of goods and services."

In this work, as in the work on depreciation and in later work, Hotelling was very concerned with the comparison of the operation of the market with "socially desirable" outcomes. Thus, in the work on mining economics, he compared the output schedules of privately owned mines with those run according to the objective of maximising social good; in the analysis of spatial

---

[52] See, for example, Ponsard (1983) who states that Hotelling's 1929 article "was to have a belated but enormous impact" (p. 106) and Dean, Leahy and McKee (1970) who stated that Hotelling's paper "has given rise to ideas and concepts that continue to cause considerable debate among spatial economic theorists" (p. vii). One of the first, and one of the most influential, commentaries upon Hotelling's work was Smithies (1941), and the Hotelling Law (that spatial competition leads to clustering of competitors) is sometimes known as the Hotelling–Smithies Law.

[53] See Lerner and Singer (1937).

competition, he was concerned that the operation of the free market would lead to a socially undesirable clustering of competitors and, as is described below, he turned his attention to such issues at a much higher level of generality in his 1938 presidential address to the Econometric Society. The theme of what constitutes the "best" for society is common to all this work.

From a chronological point of view, however, the next grouping of research to be considered is his work on the nature of supply and demand equations, research which was stimulated by a fascination with Edgeworth's taxation paradox.

V. THE NATURE OF SUPPLY AND DEMAND EQUATIONS

As was noted above, in his early career as an economist, Hotelling spent some of his time "translating" the work of earlier economists into the language of mathematics. In 1925, for example, he produced a manuscript simply entitled "The equations of classical economics", in which he presented the mathematics of deriving demand equations from a budget constrained utility maximisation framework; from this he began his investigations into the nature of demand and supply functions. In the following three years he devoted considerable time to an examination of their nature, and became particularly interested in Edgeworth's paradox.

Edgeworth's paradox concerns the possibility that the imposition of a tax on one good supplied by a monopolist who also supplies a related good (such a railway owner who supplies first- and second-class travel) may lead to a lowering, **as opposed to a raising**, of the market price of the taxed good. The paradox may be generated in an analytical framework which considers the cross-price effects in a **system** of consumer demands when a tax is raised upon one good: the analysis proceeds within **a system of interrelated markets** and thus is a departure from the partial analysis of one market alone. Edgeworth had first published this result in two papers of 1897,[54] although Hotelling's reference to Edgeworth's analysis appears to have been entirely through the *Collected Papers* of 1925.

Hotelling's work on the fundamental nature of supply and demand functions seems to have been first prompted by the belief that attempts to illustrate the paradox with numerical examples often failed because the particular demand and supply functions used did not satisfy the restrictions dictated by economic theory. In this, he was engaged not merely in a theoretical exercise, but was most concerned with the policy implications. He observed that:

> in many practical questions of governmental policy the best expert advice has gone astray because of reliance on the simplified cases treated in the textbooks, in which the correlation of demand for different commodities is neglected. (1932, p. 583).

[54] See Edgeworth (1897, 1897a and also 1899 and 1910); all are reprinted in Edgeworth (1925).

In 1928, Hotelling sketched out a numerical example of the paradox, using simple linear demand functions,[55] and showed it to Holbrook Working who was a colleague at the FRI. Working felt that Hotelling had identified a most interesting question, but that the sketch was rather too mathematical and required a non-mathematical statement of the results and their derivation.[56] Working also had two technical objections to the demand functions used in the example; these objections are cited by Hotelling in the version of the paper finally published in the *Journal of Political Economy* in 1932. Working's comments led Hotelling to provide a detailed analysis of the theoretical restrictions on the price effects in demand and supply functions, and in so doing introduced a number of then novel, and most powerful, concepts. These included, *inter alia*, the ideas of duality and integrability. On the former, it is to be noted that Samuelson, in the revised edition of his *Foundations of Economic Analysis* (1983) remarked that Hotelling's 1932 paper "was the inspiration for this book's forays into duality theory". (p. 453).

Hotelling's work, which culminated in this paper, is one of the earliest and most complete contributions to the theory of integrability; however, the paper of 1932 was directed more to the theory of the producer than to the theory of the budget constrained consumer. Thus his analysis of the consumer is essentially that of a consumer acting without a budget constraint.[57] The integrability conditions which evolve are correctly stated for the producer (who faces no budget constraint) and for the consumer in the specific circumstances described; the integrability conditions are, in this case, that the **observed** cross-price effects are symmetric. In a later published paper (1935) he addressed directly the more traditional, general, income-constrained consumer's problem and there stated the more familiar integrability conditions in the context of constrained choice.[58]

---

[55] It is interesting to note that Hotelling concentrated upon the restrictions of economic theory as they relate to price effects in demand functions; he nowhere appears to have examined the restrictions on the income effects. The linear demand functions he used did not incorporate an income term explicitly; as a consequence, they were not homogeneous of degree zero in prices and income, as dictated by economic theory.

[56] This detail is in a letter from Working to Hotelling, 3 March 1928.

[57] An alternative interpretation of Hotelling's apparently restrictive model is that the consumer has a separable utility function and one of the goods (for a proper choice of units) has a constant, unit, marginal utility. Let this function be $U(x_1, x_2, ..., x_n) = x_1 + V(x_2, ..., x_n)$. Then maximising U subject to the budget constraint $p_1 x_1 + p_2 x_2 + \cdots + p_n x_n = m$ may, by setting p_1 as numeraire, be seen as equivalent to maximising the function $V(x_2, ..., x_n) - (p_2 x_2 + \cdots + p_n x_n) + m$ which is, of course, equivalent to maximising, without constraint, $V(x_2, ..., x_n) - (p_2 x_2 + \cdots + p_n x_n)$ which is precisely the form used in Hotelling (1932). This interpretation is due to Samuelson (1950, p. 357), and it is to be noted that this interpretation requires both cardinality and separability.

[58] This condition is most succinctly stated as the requirement that the matrix of Slutsky substitution terms is symmetric and negative semi-definite; this is not the same as the condition given earlier, in that the observed price response terms and the Slutsky response terms differ by the income effect, and in general the income effects are **not** symmetric. If the utility function is

His research output on the nature of supply and demand equations, although encapsulated in only three published papers, is represented in the unpublished papers by many manuscripts and, perhaps more importantly, by a fascinating set of letters to and from Henry Schultz. During the 1930s Schultz was working on his monumental and path-breaking text *Theory and Measurement of Demand* (published in 1938), and the two men were in frequent correspondence. Hotelling provided Schultz with his integrability conditions, suggesting that they be used either to reduce the number of parameters to be estimated (and thereby increase statistical efficiency), or to test the underlying utility theory once the demand functions had been estimated freely. Many of the results obtained by Schultz contradicted the theoretical predictions of Hotelling's integrability conditions (namely that the **observed** cross-price effects, as opposed to the Slutsky substitution terms, be symmetric), and these occurrences were described by Hotelling as "a most striking phenomenon, for which some explanation is required" (1932, p. 594, fn. 6). The striking quality of the results may have been lessened had the appropriate integrability conditions been examined.[59] Hotelling had not offered Schultz the integrability conditions applicable to an income-constrained world but had offered the budget-free conditions. Hotelling was, however, fully aware of this limitation and in the same paper, of 1932, examined the necessary modifications to his integrability conditions if constrained choice were to be considered. This analysis was extended in his later paper of 1935.[60]

Both the published and unpublished Hotelling manuscripts on the topic all illustrate his keen awareness of the role of first- **and** second-order conditions for optimisation.[61] Further, in this work Hotelling introduced, albeit implicitly, the concept of the maximum value function and hence the concept of duality:

> Just as we have a utility (or profit) function *u* of the quantities consumed whose derivatives are the prices, there is, dually, a function of the prices whose derivatives are the quantities consumed. The existence of such a function, which heretofore does not seem to have been noticed, is assured by (7) [Hotelling's

homothetic then the income elasticities are all identical and are unity, in which case symmetry of the observed (Marshallian) price responses is guaranteed. However, the integrability condition may, of course, only be applied to Marshallian demands which are homogeneous of degree zero in their arguments (all prices and income) **and** satisfy "adding-up" (that is, the demands, when weighted by prices, sum to income identically).

[59] An additional problem with Schultz's empirical work is that the identification problem was, not surprisingly, addressed in an incomplete fashion.

[60] However, even in this paper, the treatment of income effects is incomplete.

[61] In Hotelling (1932), for example, he set out the sufficient second conditions for a maximum of an unconstrained function in *n* arguments, namely that the Hessian matrix should be negative definite at the stationary point, and in the context of the price-taking producer, for example, he observed that these conditions of themselves predict that the supply function is upward sloping in output price. (p. 597).

equation (7) provided the appropriate integrability condition]. On the basis of physical analogies we may call this the "price potential". (1932, p. 594).

Hotelling developed this work into a definition of "rationality": if market demand functions were to be estimated freely, then the integrability conditions could be used as a check of the existence of a utility function from which the observed demand functions could be derived. When used in this way, Hotelling (1932) suggested that:

> The difference of two symmetrically placed coefficients could be taken as a measure of the degree of inconsistency in buyers' judgements, or of the rigidity of an absolute limit on their money income expenditures.... (p. 598).

Further, in an undated, unpublished, paper (with others of 1928) he stated:

> "Rational Action" may be taken to mean a system of demand functions such that a "potential" U exists with $p_i = \delta U/\delta q_i$. Such demand and supply functions may well be taken as central, all others being treated as more or less casual deviations, often of only temporary importance.[62]

In this manuscript he also went on to suggest the implications of the theory for estimation purposes:

> The statistical determination of $\delta p_i/\delta q_j$, which equals $\delta p_j/\delta q_i$ for "rational action", involves a least-squares solution & ideas of correlation which generalise the ordinary calculus of correlation by replacing individual variables by matrices. These matrices will, moreover, be symmetric, giving rise to an interesting theory.

The ideas inherent in these passages were those which lay at the foundation of his correspondence with Schultz, and which were the basis of his advice regarding the use of the theory in estimation or testing.

Through the papers of 1932, 1933 and 1935 he developed the ideas of duality, integrability and the role of second-order conditions for the generation of economically meaningful propositions. In this, his work was contemporaneous with that of Hicks and Allen[63] (1934) and may be seen as the precursor of much current research, especially in the area of duality.

VI. WELFARE ECONOMICS

The idea of the possible divergence between optimum social outcomes and those brought about by market tendencies was the focus of Hotelling's research for much of the 1930s and was partly stimulated by his own political outlook, which was that of a market socialist. Hotelling was never prepared to accept the status quo for its own sake; this was a view he had held from his entry into economics. He once wrote:

[62] In this remark, Hotelling was working with the unlimited budget model (or equivalent model as described in footnote 57 above).

[63] See especially Allen (1932 and 1938). Hicks and Allen (1934), and Hicks (1939); for a brief historical survey of this material, see Samuelson (1950).

However, growing up, I developed deeper interests in some things that are far removed from journalism, particularly economics that might possibly impress some people to the point of making some changes in the institutions with which they had become familiar, which I was aleady prepared to condemn vigorously, ...[64]

Much of his work on welfare economics, most of it only available in the unpublished papers, culminated in his famous presidential address to the Econometric Society in 1938. However, before describing this paper, it is worth noting that during the early 1930s Hotelling wrote a number of related (unpublished) manuscripts with titles such as "Measurement of the inefficiency of sales taxes and of controlled capitalism", "The comparative disutility of income vs. excise or sales taxes", "Problems of valuation and rate-making in public enterprises", and "Prosperity through increased production".[65] These papers had a common theme, namely an examination of the role of government in economic affairs and a comparison of the outcomes arrived at through "optimal" government action and those arrived at through the operation of the market. In this the work had much in common with his earlier analyses of spatial competition and of mining economics in which similar comparisons were made. The thrust of his work on income and excise taxes cited above was to discover what characteristics an "optimal" taxation scheme would have,[66] while the paper regarding production argued for particular government intervention to alleviate the effects of the 1930s depression.[67] The paper regarding rate-making in public enterprises sought to examine the optimal pricing policy of a state owned operation.

[64] Quoted from the transcript of his acceptance speech, Rochester University, 16 May 1963, on the occasion of his receiving an honorary degree.

[65] This paper was read before the Econometric Society, 27 December 1933, but never published.

[66] In this connection it is worth noting the remark in Hotelling (1932):

If one and only one of the commodities has a supply or a demand independent of its price, ... the whole revenue should, if possible, be derived by taxing this one commodity, (p. 607).

This is effectively a Ramsey pricing conclusion.

[67] In "Prosperity through increased production" he stated that:

... by certain reorganizations which I shall briefly outline, it is possible to bring about such gains in the efficiency of the economic mechanism as to insure (sic) to the entire population a level of prosperity exceeding any in history.... It has often been said that "every tub must stand on its own bottom,".... This theory is as the bottom of a very large share of our difficulties.

He then proceeded to outline his theory of marginal-cost pricing for all industries, a theme to which he was to return in the presidential address. A related paper, "Curtailing production is anti-social", was published by *Columbia Alumni News*, and there, for example, he observed that:

The success of the government's recovery program ... must be judged, not in terms of price levels, but in terms of the physical goods and services which are put into the hands of consumers ... the chief thing needed is to increase physical production. In this respect much that is being done at Washington is definitely in the wrong direction. The attempts to increase the prices and curtail the production of oil, agricultural products, and other commodities are anti-social. (p. 3).

Hotelling's work on welfare economics thus proceeded over a number of years, and culminated in his famous address to the Econometric Society in 1938: "The general welfare in relation to problems of taxation and of railway and utility rates". This paper provided a statement of the "marginal-cost pricing principle" itself and, through the extensive use of mathematics, gave a very detailed explanation of the implications of the principle in practice. It is recognised as one of the fundamental contributions to welfare economics and the marginal-cost pricing controversy in particular.[68] In it, Hotelling advanced a number of policy prescriptions; one of the most important is that all industries should price at marginal cost and, if this implies that some state run industries make losses (due to their having decreasing average cost schedules), then such losses should be subsidised out of lump-sum taxes. He provided a mathematical derivation of this proposition within a general equilibrium framework and discussed the kinds of taxes which would satisfy the requirement that welfare was maximised (or equivalently that the losses due to taxation be minimised). In identifying the set of appropriate lump-sum taxes, Hotelling made some errors, particularly in his suggestion that some income taxes would satisfy the requirement. Nevertheless, such slips do not detract from the path-breaking nature of this work.

Of all his published work, this was, perhaps, the most immediately influential within the academic community. It was also a most sophisticated application of advanced mathematical tools in economics, making use of line integrals to generalise the concepts of producers' and consumers' surplus for many commodities. What gave this paper its immediate appeal was its readily understood theoretical and policy conclusions, notwithstanding their mathematical derivation and controversial subject matter.

This paper derived from the earlier manuscripts cited above, and much of the preparatory analysis can be found in those papers. For example, in the papers on income and excise taxes, Hotelling showed, through a mathematical analysis of indifference curves, that an individual's welfare is greater if taxes are raised through lump-sum levies rather than raising the same amount through excise taxes. This follows because no marginal conditions for equilibrium are disturbed by a lump-sum tax, whereas an excise tax drives a "wedge" between producer and consumer prices which then leads to a contravention of the necessary marginal conditions for a Pareto optimum. This proposition is central to the thesis.[69]

Regarding the role of government, there is a passage in his paper "Prosperity

---

[68] For an historical overview of welfare economics see, for example, Herbert and Ekelund (1984); there it is observed that:

> Much of the economic analysis stimulated by Hotelling's contribution, generalizing Dupuit's approach and blending it with the Paretian welfare tradition, subsequently developed in the context of public goods.... (p. 64).

[69] Of course, the marginal conditions are only undisturbed if the lump-sum tax applies to all individuals—if it only applies to those in receipt of earned income then the real return to work has been disturbed, and there is likely to be a disincentive effect to work which will alter the choice between work and leisure.

through increased production" (presented to the Econometric Society in December 1933) which is worth quoting at length. In it Hotelling argues the convincing case for the use of mathematics in economics (while also arguing against laissez-faire government):

> A famous proposition of classical economics, an idea which in the popular mind is chiefly associated with Adam Smith and God, sets forth that, in a certain sense, the total of satisfactions is a maximum when there is no governmental interference with business, either in the form of subsidies or of taxes, but each individual, and by inference each corporation, is left free to decide how much or little it or he will produce or sell or buy, and of what; provided only, however, that each acts as accurately to maximise his own gain. The conclusion is far more popular than the reasoning which leads to it. By accepting the conclusion without troubling about a formal demonstration, multitudes of people have lost sight of the limitations of the proposition. It has often been said that mathematical operations are dangerous, as tending to induce a false sense of confidence in their results because the mathematical symbols obscure the underlying relations. On the contrary, however, mathematical demonstrations, which are to be distinguished from mere calculations, serve a very useful purpose in forcing into the open assumptions which otherwise are accepted tacitly, and thus to emphasise the limitations of the result. Thus, in contrast with the usual notion, one who has written out a mathematical demonstration of a proposition, whether in economics or physics, is likely to have less confidence in the applicability of the conclusion to an actual situation than has a person who knows the conclusion, but not the reasoning on which it is based.
>
> In asserting that a maximum of total satisfactions is to be reached by laissez-faire, it is usually forgotten that the proof of the statement calls for competition of a type not found in large-scale modern industry.

This work demonstrated several aspcts of Hotelling's outlook. One was his commitment to the erection of a rigorous theoretical framework for economics, a commitment which for him required the extensive use of mathematics; second was his strong interest in policy issues and third was his belief in market socialism. These three characteristics are evident to a most marked degree in his welfare economics. As noted above, his presidential address proved to be controversial, and Frisch (1939 and 1939a) replied to it in *Econometrica*, and a debate may be found there between Hotelling and Frisch. Frisch raised a number of objections to Hotelling's thesis; some were effectively answered by Hotelling, others not, and on one point both were wrong.[70] The debate was not confined to these two discussants, however, and many others entered the fray.[71]

---

[70] As a technical point, the difficulties with Hotelling's work on welfare economics stem from his rather too strict a dichotomisation between producers and consumers. His work falls into the two categories of the analysis of optimal production conditions and the analysis of optimal exchange conditions. As Samuelson (1983) noted, "the two treatments are never adequately integrated." (p. 218). The point upon which both were wrong was Frisch's suggestion that prices need only be **proportional** to marginal cost; Hotelling agreed with this, but if **all** prices are proportional to marginal costs, then the factor of proportionality must be unity!

[71] Ruggles (1949–50), has provided an effective description of the debate. See also Morrison and Pfouts (1981).

VII. Conclusion

Hotelling's publications in economics provide a set of seminal contributions to a number of disparate areas of our discipline. Each illustrates a sound application of mathematical analysis to important economic issues and, even though most of his papers were not immediately recognised nor were immediately influential, they have now taken on the role of fundamental contributions. However, his work on both spatial competition and on welfare economics did have immediate impact and has formed the basis of a literature which dates from their publication.

Hotelling was himself sensitive to the fact that his publications in the fields of depreciation and supply and demand functions had had little contemporary impact. For example, in 1959 Carl Shoup (of Columbia University) wrote to Hotelling informing him that he had made reference to the 1932 paper on Edgeworth's paradox and Hotelling replied:

> I have had the impression that few people ever read my 1932 paper and it is good to know that you, at least, think it worth taking seriously. Each of my two papers in the *Journal of Political Economy*, the 1931 article on mineral economics and the 1932 to which you allude, represented about a quarter of my time for several years and it is a pleasure to know that the work was not in vain.

Clearly, then, Hotelling had spent a great deal of time on these two publications, and was, naturally, aware that both papers had had little impact upon economic research.[72] This may be explained by at least two observations. First, all of Hotelling's work requires the reader to have a mastery of mathematical analysis, and the profession of economists was, particularly at the time of his publications, ill-prepared for the sophisticated use of mathematics in economics; second, the issues associated with the economics of depression overtook many other concerns, and especially dominated concern about the finite nature of some natural resources. For example, it was not until issues of growth economics, and especially the issues of the limits to growth, became fashionable in the 1960s that economists "re-discovered" Hotelling's seminal contribution of some thirty years earlier. His work on demand and supply functions is yet to receive the acclaim it deserves.

The depression was itself the source of inspiration for some of Hotelling's own work, notably that concerned with production levels and what he saw as the potential failure of laissez-faire government policy, research which was based on principles expounded to such good effect in his presidential address.

His influence upon contemporary research through his few publications is great, yet behind those papers lies a wealth of manuscripts in which the ideas may be seen to have evolved and been developed. They provide evidence that

---

[72] Indeed, the literature concerned with irreplaceable assets took no steps forward until the middle 1960s; theoretical work following Hotelling (1931) first began to appear in the 1960s. See, for example, Herfindahl (1967), Gordon (1967) and Cummings (1969). Also, the first edition of Samuelson's *Foundations of Economic Analysis* does not acknowledge the debt to Hotelling (1932), and his work in the area of demand and supply is little known or appreciated to this day.

his research was neither encapsulated nor confined to his publications and will provide economists with a rich source of primary material upon which to base their research. His stature amongst economists in great and this will only be enhanced by a reading of his currently unpublished papers and correspondence.

Hotelling was a leading economist of his day and also a highly revered teacher; although he published very few papers in economic theory, his contributions have each now assumed the stature of classics. A particular feature of Hotelling's economics was the use of highly sophisticated mathematical techniques which enabled him to derive results which had, hitherto, been inaccessible.[73] Most significantly, he published in diverse fields demonstrating that mathematical methods are a most powerful tool with which to analyse many different issues in economics.

To date there has been no systematic study of Hotelling's life and work; this is a major omission in the study of the history of economics. His place in the development of economics, statistics, and the applications of mathematics to the social sciences has never been the subject of study, yet there are few figures who have displayed such mastery in so many different fields, and fewer still whose publication record bears comparison. Importantly, those areas of economics into which Hotelling conducted research have not diminished in significance from the time of his own contributions; issues of mining economics, taxation, welfare, utility and demand are still major concerns of the profession. That his work should today be seen as the starting point for research in so many areas, and his mathematical and statistical methods be so widely adopted, is a great testimony to its worth, yet our ignorance of the author frustrates a full understanding of his work and its true importance.

This introductory chapter is only a first, tentative, step in identifying the range and importance of Hotelling's contributions to economics; further work will facilitate deeper appreciation of Harold Hotelling, his work and, in consequence, those areas of economics in which he pursued his research.

The Department of Economics   ADRIAN C. DARNELL
University of Durham
June 1988

REFERENCES

Allen, R. G. D. (1931): Review of *Mathematical Introduction to Economics. Economica*, **11**, 108–9.
———— (1932): The foundation of a mathematical theory of exchange. *Economica*, **12**, 197–226.

[73] This was, as stressed above, no accident. From his Rochester acceptance speech, 1963, he observed that on leaving journalism:

> I couldn't escape the lure of mathematics which was so extremely beautiful and it was hoped and appeared that the mathematics would be extremely prolific of new economic theorems and ideas if followed out.

Allen, R. G. D. (1938): *Mathematical Analysis for Economists*. London: Macmillan.

American Economic Review (1974): In Memoriam, Harold Hotelling. **64**, 1102–3.

Anscombe, F. J. (1980): Comment. *Journal of the American Statistical Association*, **75**, 331.

Arrow, K. J. (1987): Harold Hotelling, in Eatwell, J, Milgate, M. and Newman, P. (eds.), *The New Palgrave*. London: Macmillan, 2, pp. 670–1.

d'Aspremont, P., Gabszewicz, J.-J. and Thisse, J.-T. (1979): On Hotelling's "Stability in competition". *Econometrica*, **47**, 1145–56.

Bell, E. T. (1937): *Men of Mathematics*, New York: Simon and Schuster.

Bowley, A. L. (1932): Review of *Mathematical Introduction to Economics*. *Economic Journal*, **42**, 93–4.

Conrad, J. M. and Clark, C. W. (1987): *Natural Resource Economics: Notes and Problems*. Cambridge: Cambridge University Press.

Cummings, R. G. (1969): Some extensions of the economic theory of exhaustible resources. *Western Economic Journal*, 7, 201–10.

Darnell A. C. (1982): Bowley, Wicksell and the development of mathematical economics. *Scottish Journal of Political Economy*, **29**, 156–80.

———— (1988): Harold Hotelling 1895–1973. *Statistical Science*, 3, 57–62.

Dean, R. D., Leahy, W. H., and McKee, D. L. (1970): *Spatial Economic Theory*. New York: The Free Press.

Devarajan, S. and Fisher, A. C. (1981): Hotelling's "Economics of exhaustible resources": Fifty years later. *Journal of Economic Literature*, **19**, 65–73.

Edgeworth, F. Y. (1897): Teoria pura del monopolio. *Giornale degli Economisti*. Translated and reprinted as "The theory of monopoly" in Edgeworth (1925), Vol. I, pp. 111–42.

———— (1897a): The pure theory of taxation. *Economic Journal*, 7, 46–70, 226–38 and 550–71; reprinted in Edgeworth (1925), Vol. II, pp. 63–125.

———— (1899): Professor Seligman on the mathematical method in political economy. *Economic Journal*, 9, 286–315; reprinted as "Professor Seligman on the theory of monopoly" in Edgeworth (1925), Vol. I, pp. 143–71.

———— (1910): Applications of probabilities to economics. *Economic Journal*, **20**, 284–304 and 441–65; reprinted in Edgeworth (1925), Vol. II, pp. 387–428.

———— (1925): *Papers Relating to Political Economy*, Volumes I, II and III. London: Macmillan.

Evans, G. C. (1922): A simple theory of competition. *American Mathematical Monthly*, **29**, 371–80.

———— (1924): The dynamics of monopoly. *American Mathematical Monthly*, **31**, 77–83.

———— (1925): The mathematical theory of economics. *American Mathematical Monthly*, **32**, 104–10.

———— (1925a): Economics and the calculus of variations. *Proceedings of the National Academy of Science*, **11**, 90–5.

———— (1930): *Mathematical Introduction to Economics*. New York: McGraw-Hill.

Frisch, R. (1939): The Dupuit taxation theorem. *Econometrica*, 7, 145–50.

———— (1939a): A further note on the Dupuit taxation theorem. *Econometrica*, 7, 156–7.

Gordon, R. L. (1967): A reinterpretation of the pure theory of exhaustion. *Journal of Political Economy*, **75**, 274–86.

Gray, L. C. (1914): Rent under the assumption of exhaustibility. *Quarterly Journal of Economics*, **28**, 497–519.

Grubel, H. G. and Boland, L. A. (1986): On the efficient use of mathematics in economics: Some theory, facts and results of an opinion survey. *Kyklos*, **39**, 419–42.

Herbert, R. F. and Ekelund, R. F. (1984): Welfare Economics in Creedy, J. and O'Brien, D. P. (eds.), *Economic Analysis in Historical Perspective*. London: Butterworths, pp. 46–83.

Herfindahl, O. C. (1967): Depletion and economic theory, Chapter 3 in Gaffney, M. (ed.), *Extractive Resources and Taxation*. Madison: University of Wisconsin Press, pp. 63–90.

Hicks, J. R. (1939): *Value and Capital*. Oxford: Oxford University Press.

Hicks, J. R. and Allen, R. G. D. (1934): A reconsideration of the theory of value, part I. *Economica*, n.s., **1**, 52–76 and A reconsideration of the theory of value, part II: A mathematical theory of individual demand functions. *Economica*, n.s., **1**, 196–219.

Hotelling, H. (1925): A general mathematical theory of depreciation. *Journal of the American Statistical Association*, **20**, 340–53.

———— (1925a): Economics problems involving maxima of functionals. *Bulletin of the American Mathematical Society*, **31**, 202.

———— (1925b): Theory of mine economics. *Bulletin of the American Mathematical Society*, **31**, 389.

———— (1925c): Three-dimensional manifolds of states of motion. *Transactions of the American Mathematical Society*, **27**, 329–44.

———— (1925d): The distribution of correlation ratios calculated from random data. *Proceedings of the National Academy of Science*, **11**, 657–62.

———— (1926): Multiple-sheeted spaces and manifolds of states of motion *Transactions of the American Mathematical Society*, **38**, 479–90.

———— (1926a): Two generalisations of the Pearsonian correlation coefficient. *Bulletin of the American Mathematical Society*, **32**, 98.

———— (1927): Differential equations subject to error, and population estimates. *Journal of the American Statistical Society*, **22**, 283–314.

———— (1929): Stability in competition. *Economic Journal*, **39**, 41–57.

———— (1930): The consistency and ultimate distribution of optimum statistics. *Transactions of the American Mathematical Society*, **32**, 847–59.

———— (1931): The economics of exhaustible resources. *Journal of Political Economy*, **39**, 137–75.

———— (1931a): The generalization of Student's ratio. *Annals of Mathematical Statistics*, **2**, 360–78.

———— (1931b): Review of *Mathematical Introduction to Economics*. *Journal of Political Economy*, **39**, 107–9.

———— (1931c): Review of *Mathematical Introduction to Economics*. *American Mathematical Monthly*, **38**, 101–3.

———— (1932): Edgeworth's taxation paradox and the nature of supply and demand functions. *Journal of Political Economy*, **40**, 577–616.

Hotelling, H. (1933): Note on Edgeworth's taxation phenomenon and Professor Garver's additional condition on demand functions. *Econometrica*, **1**, 408–9.

———— (1933a): Analysis of a complex of statistical variables with principal components. *Journal of Educational Psychology*, **24**, 417–441 and 498–520.

———— (1935): Demand functions with limited budgets. *Econometrica*, **3**, 66–78.

———— (1936): Curtailing production is anti-social. *Columbia Alumni News*, **28**, 3 and 16.

———— (1936a): Relations between two sets of variates. *Biometrika*, **28**, 321–77.

———— (1938): The general welfare in relation to problems of taxation and of railway and utility rates. *Econometrica*, **6**, 242–69.

———— (1939): The relation of prices to marginal cost in an optimum system. *Econometrica*, **7**, 151–5.

———— (1939a): A final note. *Econometrica*, **7**, 158–60.

———— (1939b): The work of Henry Schultz, *Econometrica*, **7**, 97–103.

———— (1940): The teaching of statistics. *Annals of Mathematical Statistics*, **11**, 457–70.

———— (1940a): Appendix: Questions of Preinreich. *Econometrica*, **8**, 39–44.

———— (1943): Income tax revision as proposed by Irving Fisher. *Econometrica*, **11**, 83- 7.

———— (1947): Multivariate quality control, illustrated by the air testing of sample bomb-sights. Chapter 3 in Eisenhart, C., Hastay, M. W. and Wallis, W. A. (eds.). *Selected Techniques of Statistical Analysis*. New York: McGraw-Hill, pp. 111–84.

———— (1955): Econometrics. *Encyclopedia Americana*, Chicago: Americana Corp., p. 556.

———— (1958): C. F. Roos, Econometrician and mathematician. *Science*, **128**, 1194–5.

———— and Working, H. (1929): Applications of the theory of error to the interpretation of trends. *Journal of the American Statistical Association*, **24**, 73–85.

Kruskal, W. H. (1980): Comment—first interaction with Harold Hotelling; testing the Norden bomb-sight. *Journal of the American Statistical Association*, **75**, 331–3.

Lerner, A. P. and Singer, H. W. (1937): Some notes on duopoly and spatial competition. *Journal of Political Economy*, **45**, 145–86.

Madow, W. G. (1960): Harold Hotelling in Olkin, I., Ghurye, S. G., Hoeffding, W., Madow, W. G. and Mann, H. B. (eds.), *Contributions to Probability and Statistics, Essays in Honour of Harold Hotelling*. Stanford: Stanford University Press, pp. 3- 5.

Morrison, C. C. and Pfouts, R. W. (1981): Hotelling's proof of the marginal cost pricing theorem. *Atlantic Economic Journal*, **9**, 34–7.

Neyman, J. (1960): Harold Hotelling—A leader in mathematical statistics, in Olkin, I., Ghurye, S. G., Hoeffding, W., Madow, W. G. and Mann, H. B. (eds.), *Contributions to Probability and Statistics, Essays in Honour of Harold Hotelling*. Stanford: Stanford University Press, pp. 6–10.

Olkin, I, Ghurye, S. G., Hoeffding, W., Madow, W. G. and Mann, H. B. (eds.), (1960): *Contributions to Probability and Statistics, Essays in Honour of Harold Hotelling*. Stanford: Stanford University Press.

Peterson, F. M. and Fisher, A. C. (1977): The exploitation of exhaustible resources: A survey. *Economic Journal*, **87**, 681–721.

Pfouts, R. W. and Leadbetter, M. R. (1979): Hotelling, Harold, *The International Encyclopaedia of the Social Sciences*, **18**, 325–8.

Ponsard, C. (1983): *History of Spatial Economic Theory*. Berlin: Springer-Verlag. (This is a revised, translated, version of the original French edition of 1958.)

Preinreich, G. A. D. (1940): The economic life of industrial equipment (with Appendix). *Econometrica*, **8**, 12–39 and 39–44.

Ramsey, F. P. (1927): A contribution to the theory of taxation. *Economic Journal*, **37**, 47–61.

———— (1928): A mathematical theory of saving. *Economic Journal*, **39**, 543–59.

Ruggles, N. (1949–50): Recent development in the theory of marginal-cost pricing, *Review of Economic Studies*, **17**, 107–26.

Samuelson, P. A. (1950): The problem of integrability in utility theory. *Economica*, n.s., **17**, 355–85.

———— (1960): Harold Hotelling as a mathematical economist. *American Statistician*, **14**, 21–5.

———— (1983): *Foundations of Economic Analysis*. Cambridge, MA, Harvard Economic Studies, Vol. 80 (revised edition).

Schultz, H. (1938): *The Theory and Measurement of Demand*. Chicago: Chicago University Press.

Smith, W. L. (1978): Harold Hotelling 1895–1973. *The Annals of Statistics*, **6**, 1173–83.

Smithies, A (1941): Optimum location in spatial competition. *Journal of Political Economy*, **49**, 423–39.

Sraffa, P. (1926): The laws of returns under competitive conditions. *Economic Journal*, **36**, 535–50.

Taylor, J. S. (1923): A statistical theory of depreciation based on unit cost. *The Journal of the American Statistical Association*, **18**, 1010–23.

———— (1925): A note on the theory of depreciation. *Bulletin of the American Mathematical Society*, **31**, 222.

Wallis, W. A. (1980): The Statistical Research Group, 1942–45. *Journal of the American Statistical Association*, **75**, 320–30 and 334–5.

Wright, P. G. (1931): Review of *Mathematical Introduction to Economics*. *American Economic Review*, **21**, 93–5.

Bibliography of Harold Hotelling

1. Publications Other Than Reviews, Reprints and Translations

1925

Three-dimensional manifolds of states of motion. *Transactions of the American Mathematical Society*, **27**, 329–44.

A general mathematical theory of depreciation. *Journal of the American Statistical Association*, **20**, 340–53.

The distribution of correlation ratios calculated from random data. *Proceedings of the National Academy of Science*, **11**, 657–62.

1926

Multiple-sheeted spaces and manifolds of states of motion. *Transactions of the American Mathematical Society*, **28**, 479–90.

1927

An application of analysis situs to statistics. *Bulletin of the American Mathematical Society*, **33**, 467–76.

Differential equations subject to error, and population estimates, *Journal of the American Statistical Association*, **22**, 283–314.

1928

Spaces of statistics and their metrization. *Science*, **67**, 149–50.

The physical state of protoplasm (with L. G. M. Baas Becking and Henriette van der Sande Bakhuysen). *Verhandel. Koninkl. Akad. Wetenschap. Amsterdam, Afdeeling Natuurkunde* (Tweedie Sect.), **25**(5), 28–31.

1929

Applications of the theory of error to the interpretation of trends (with Holbrook Working). *Journal of the American Statistical Association*, **24**, 73–85.

Stability in competition. *Economic Journal*, **39**, 41–57.

A determinant game. *American Mathematical Monthly*, **36**, 285.

1930

British statistics and statisticians today. *Journal of the American Statistical Association*, **25**, 186–90.

The consistency and ultimate distribution of optimum statistics. *Transactions of the American Mathematical Society*, **32**, 847–59.

1931

Recent improvements in statistical inference. *Journal of the American Statistical Association*, **26**, 79–89.

The economics of exhaustible resources. *Journal of Political Economy*, **39**, 137–75.

Frequency distributions. *Encyclopaedia of the Social Sciences*, Volume 6. New York: Macmillan, pp. 484–9.

Causes of birth rate fluctuations (with Floy Hotelling). *Journal of the American Statistical Association*, **26**, 135–49.

The generalization of Student's ratio. *Annals of Mathematical Statistics*, **2**, 360–78.

1932

The limits of a measure of skewness (with Leonard M. Solomons). *Annals of Mathematical Statistics*, **3**, 141–2.

A new analysis of duration of pregnancy data (with Floy Hotelling). *American Journal of Obstetrics and Gynaecology*, **23**, 643–57.

Edgeworth's taxation paradox and the nature of demand and supply functions. *Journal of Political Economy*, **40**, 577–616.

1933

Analysis of a complex of statistical variables into principal components. *Journal of Educational Psychology*, **24**, 417–41 and 498–520.

Note on Edgeworth's taxation phenomenon and Professor Garver's additional condition on demand functions. *Econometrica*, **1**, 408–9.

1935

Demand functions with limited budgets. *Econometrica*, **3**, 66–78.

The most predictable criterion. *Journal of Educational Psychology*, **26**, 139–42.

1936

Rank correlation and tests of significance involving no assumption of normality (with Margaret Richards Pabst). *Annals of Mathematical Statistics*, **7**, 29–43.

Measurements and correlations of pelves of Indians of the Southwest (with W. W. Howells). *American Journal of Physical Anthropology*, **21**, 91–106.

Some little-known applications of mathematics. *Mathematics Teacher*, **29**, 157–69.

Simplified calculation of principal components. *Psychometrika*, **1**, 27–35.

Relations between two sets of variates, *Biometrika*, **28**, 321–77.

Curtailing production is anti-social. *Columbia Alumni News*, **28**(4), 3 and 16.

1937

Effects of pregnancy, mouth acidity and age on dental caries (with Daniel E. Ziskin). *Journal of Dental Research*, **16**, 507–19.

1938

The transformation of statistics to simplify their distribution (with Lester R. Frankel). *Annals of Mathematical Statistics*, **9**, 87–96.

The general welfare in relation to problems of taxation and of railway and utility rates. *Econometrica*, **6**, 242–69.

1939

Tubes and spheres in *n*-spaces and a class of statistical problems. *American Journal of Mathematics*, **61**, 440–60.

The work of Henry Schultz. *Econometrica*, **7**, 97–103.

The relation of prices to marginal cost in an optimum system. *Econometrica*, **7**, 151–5.

A final note. *Econometrica*, **7**, 158–60.

1940

The selection of variates for use in prediction with some comments on the general problem of nuisance parameters. *Annals of Mathematical Statistics*, **11**, 271–83.

The teaching of statistics. *Annals of Mathematical Statistics*, **111**, 457–70.

1941

Experimental determination of the maximum of a function. *Annals of Mathematical Statistics*, **12**, 20–45.

Presidential Address (to Indian Statistical Congress at Madras). *Sankhya*, **5**, 127–9.

1942

Problems of prediction. *American Journal of Sociology*, **48**, 61–76.

Rotations in psychology and the statistical revolution. *Science*, **95**, 504–7.

Foreword to *Tables of Probability Functions*, Vol. II. National Bureau of Standards, Washington, DC: Government Printing Office.

1943

Income-tax revision as proposed by Irving Fisher. *Econometrica*, **11**, 83–7.

Some new methods in matrix calculation. *Annals of Mathematical Statistics*, **14**, 1–34.

Dr. Peters' criticism of Fisher's statistics. *Journal of Educational Research*, **36**, 707–11.

Further points on matrix calculation and simultaneous equations. *Annals of Mathematical Statistics*, **14**, 440–1.

1944

Some improvements in weighing and other experimental techniques. *Annals of Mathematical Statistics*, **15**, 297–306.

Note on a matric theorem of A. T. Craig. *Annals of Mathematical Statistics*, **15**, 427–9.

1945

Graduate work in statistics at Columbia University. *Biometrics Bulletin*, **1**, 22–3.

1947

The training of statisticians. *American Statistician*, **1**(3), 8–9.

Multivariate quality control, illustrated by the air testing of sample bomb-sights. Chapter 3 in Eisenhart, C., Hastay, M. W. and Wallis, W. A., (eds.) *Selected Techniques of Statistical Analysis*. New York: McGraw-Hill, pp. 111–84.

Effects of non-normality at high significance levels. *Annals of Mathematical Statistics*, **18**, 608–9.

1948

Points in the report of the Institute of Mathematical Statistics Committee on the teaching of statistics. *Mathematics Teacher*, **41**, 76–7.

The teaching of statistics. A report of the Institute of Mathematical Statistics Committee on the teaching of statistics. Harold Hotelling, Chairman: Walter Bartky; W. Edwards Deming; Milton Friedman; Paul Hoel. *Annals of Mathematical Statistics*, **19**, 95–115.

Mathematical statistics and econometrics at the University of North Carolina. *Econometrica*, **16**, 125–6.

Discussion of paper of Georges Teissier. *Biometrics*, **4**, 50.

Wake Forest for future leaders. *The News and Observer*, Raleigh, NC, August 31, 1948, p. 4.

Mathematics in our schools. *The News and Observer*, Raleigh, NC, September, 2, 1948, p. 4.

1949

Discussion of paper of René Roy. *Econometrica*, **17**, 188–9.

Stochastic processes. (Remarks by the chairman at the joint meeting of the Econometric Society and the Institute of Mathematical Statistics.) *Econometrica*, **17**, 66–8.

Practical problems of matrix calculation, *Proceedings of the Berkeley Symposium on Mathematical Statistics and Probability.* Berkeley, CA: University of California Press, pp. 275–94.

The place of statistics in the university. *Proceedings of the Berkeley Symposium on Mathematical Statistics and Probability.* Berkeley, CA: University of California Press, pp. 21–40.

1950

Some computational devices. Chapter 10 in Koopmans, T. C. (ed.), *Statistical Inference in Dynamic Economic Models.* Cowles Commission Monograph No. 10, New York: Wiley.

Needed steps in the mathematical training of social scientists. *Econometrica*, **18**, 198–9.

Why we have the two-party system. *American Journal of Economics and Sociology*, **10**, 13–5.

Comments on the report, "Furtherance of statistical education." *Revue de l'Institut Internationale de Statistique*, **18**, 69–70.

1951

The impact of R. A. Fisher on statistics. *Jornal of the American Statistical Association*, **46**, 35–46.

Abraham Wald. *American Statistician*, **5**, 18–9.

A generalized *T* test and measure of multivariate dispersion. *Proceedings of the Second Berkeley Symposium on Mathematical Statistics and Probability.* Neyman, J. (ed.), Berkeley, CA: University of California Press, pp. 23–42.

Foreword to "The Jacobians of certain matrix transformations useful in multivariate analysis." *Biometrika*, **38**, 345–6.

Discussion of the paper of J. Neyman. *Proceedings of the International Statistical Conferences at Washington*, III A. The Hague: International Statistical Institute, p. 432.

1953

New light on the correlation coefficient and its transforms. *Journal of the Royal Statistical Society*, Ser. B, **15**, 193–225, with discussion, 225–32.

1954

Multivariate analysis. Chapter 4 in Kempthorne, O., Bancroft, T. A., Gowen, J. W. and Lush, J. L. (eds.), *Statistics and Mathematics in Biology*. Ames, Iowa: Iowa State College Press.

Discussion of paper of J. Durbin, *Econometrica*, **20**, 102, 105 and 108–9.

1955

Econometrics. Encyclopedia Americana, Vol. 9. Chicago: Americana Corp., p. 556.

The correlation coefficient and its transforms—a new look. *Résumés des conférences faites au séminaire de statisque tenu à Rome, Septembre 1953*. The Hague: International Statistical Institute.

The moments of the sample median (with John T. Chu). *Annals of Mathematical Statistics*, **26**, 593–606.

Multivariate methods in testing complex equipment. *Papers Presented at the Colloquium in Statistical Design of Laboratory Experiments*. U.S. Naval Ordnance Laboratory, White Oak, MD, pp. 84–91.

1956

Les rapports entre les méthodes statistiques récentes portant sur des variables multiples et l'analyse factorielle. *L'analyse factorielle et ses applications*. Paris: Centre national de la recherche scientifique, pp. 107–19, with discussion, pp. 120–5.

1957

The relations of the newer multivariate statistical methods to factor analysis. *British Journal of Statistical Psychology*, **10**(II), 69–79.

1958

C. F. Roos, Econometrician and mathematician. *Science*, **128**, 1194–5.

The statistical method and the philosophy of science. *American Statistician*, **12**(5), 9–14.

2. REVIEWS

1926

Day, E. E.: *Statistical Analysis. Journal of the American Statistical Association*, **21**, 360–3.

Dublin, L. I.: *et al.: Population Problems in the United States and Canada. Journal of the American Statistical Association*, **21**, 503–5.

1927

Fisher, R. A.: *Statistical Methods for Research Workers. Journal of the American Statistical Association*, **22**, 411–2.

34   Harold Hotelling

1928

Elderton, W. P.: *Frequency Curves and Correlation. Bulletin of the American Mathematical Society*, **34**, 515–6.

Fisher, R. A.: *Statistical Methods for Research Workers*, 2nd ed. *Journal of the American Statistical Association*, **23**, 346.

Cournot, A.: *Researches into the Mathematical Principles of the Theory of Wealth. American Mathematical Monthly*, **35**, 439–40.

1930

Fisher, R. A.: *Statistical Methods for Research Workers*, 3rd ed. *Journal of the American Statistical Association*, **25**, 381–2.

1931

Evans, G. C.: *Mathematical Introduction to Economics. Journal of Political Economy*, **39**, 107–9.

Evans, G. C.: *Mathematical Introduction to Economics. American Mathematical Monthly*, **38**, 101–3.

1932

Shewhart, W. A.: *Economic Control of Quality of Manufactured Product. Journal of the American Statistical Association*, **27**, 215–7.

Frisch, R.: *New Methods of Measuring Marginal Utility. Journal of the American Statistical Association*, **27**, 451–2.

1933

Fisher, R. A.: *Statistical Methods for Research Workers*, 4th ed. *Journal of the American Statistical Association*, **28**, 374–5.

Secrist, H.: *The Triumph of Mediocrity in Business. Journal of the American Statistical Association*, **28**, 463–5.

1934

Reply to Secrist. *Journal of the American Statistical Association*, **29**, 198–9.

1935

Smith, J. G.: *Elementary Statistics, an Introduction to the Principles of Scientific Method. American Mathematical Monthly*, **42**, 169–71.

Snedecor, G. W.: *Calculation and Interpretation of Analysis of Variance and Covariance. Journal of the American Statistical Association*, **30**, 117–8.

Fisher, R. A.: *Statistical Methods for Research Workers*, 5th ed. *Journal of the American Statistical Association*, **30**, 118.

Roos, C. F.: *Dynamic Economics. Journal of the American Statistical Association*, **30**, 480–3.

Fisher, R. A.: *The Design of Experiments. Journal of the American Statistical Association*, **30**, 771–2.

1937

Fisher, R. A.: *Statistical Methods for Research Workers*, 6th ed. *Journal of the American Statistical Association*, **32**, 218–9.

Fisher, R. A.: *The Design of Experiments*, 2nd ed. *Journal of the American Statistical Association*, **32**, 580–2.

1938

Schultz, H.: *The Theory and Measurement of Demand. Journal of the American Statistical Association*, **33**, 744–7.

Fisher, R. A. and Yates, F.: *Statistical Tables for Biological, Agricultural and Medical Research. Science*, **88**, 596–7.

1939

Allen, R. G. D.: *Mathematical Analysis for Economists. Annals of the American Academy of Political and Social Science*, **205**, 154–5.

Fisher, R. A.: *Statistical Methods for Research Workers*, 7th ed. *Journal of the American Statistical Association*, **34**, 423 -4.

1941

Preinreich, G. A. D.: *The Present Status of Renewal Theory. Journal of Political Economy*, **49**, 136–7.

1942

Holzinger, K. L. and Harman, H. H.: *Factor Analysis. Science*, **95**, 504–7.

Horst, P. *et al.*: *The Prediction of Personal Adjustment. American Journal of Sociology*, **48**, 61–76.

Davis, H. T.: *The Analysis of Economic Time Series. Annals of the American Academy of Political and Social Science*, **221**, 219–21.

1943

Molina, E. C.: *Poisson's Exponential Binomial Limit. Journal of Educational Psychology*, **34**, 124–5.

1944

Nanavati, M. B. and Vakil, C. N (eds.): *India Speaking* (from *Annals of the American Academy of Political and Social Science*). *Asia*, **47**, 382–4.

1945

Kendall, M. G.: *The Advanced Theory of Statistics*, Vol. 1. *Bulletin of the American Mathematics Society*, **51**, 214–6.

1947

Cramer, H.: *Mathematical Methods of Statistics. Journal of Political Economy*, **55**, 393–5.

1948

Kendall, M. G.: *The Advanced Theory of Statistics*, Vol. II. *Bulletin of the American Mathematics Society*, **54**, 863–8.

A General Mathematical Theory of Depreciation[1]

In the older treatments of depreciation the cost, or "theoretical selling price" of the product of a machine, was conceived of as determined causally by the addition of a number of items of which depreciation is one. In other words, depreciation was first computed by some rather arbitrary formula not involving the theoretical selling price, which was then found by the addition of depreciation to operating costs and division by quantity of output. It will be shown in this paper that depreciation and theoretical selling price must be computed simultaneously from a pair of equations which are frequently a bit complicated. The differences in the results obtained from the arbitrary and mathematical formulae are often very large.

The simple methods referred to, which still prevail generally in business, are analogous to the naïve type of economic thought for which the only determiner of price is cost and which fails to consider the equally important rôle played by demand.

The "unit cost theory" gave the first recognition to the reciprocal relation existing between the value of a machine and the value of its product. It also deserves the credit for taking into account operating costs and output which, as in the present paper, are assumed to have been determined from experience. However, this theory in its usual form involves a number of serious errors of reasoning which have been pointed out by Dr. J. S. Taylor.[2]

As an improvement on the unit cost theory Dr. Taylor puts forward a method which, it is fair to say, is the only one that has been proposed which ever gives correct results. To it the present paper owes much. He assumes that unit cost plus (defined as operating cost, plus depreciation, plus interest on the value of the machine, all divided by the number of units of output) is to be determined by the conditions that

(a) it is either to remain constant during the machine's life or is to increase in a specified manner; and
(b) it is to be made a minimum.

---

[1] Presented before the American Mathematical Society (Chicago meeting), December 26, 1924.

[2] "A Statistical Theory of Depreciation Based on Unit Cost," *Journal of the American Statistical Association*, December, 1923, pp. 1010–1023.

This procedure is open to a possible criticism which he does not anticipate, namely, that, since the unit cost plus depends on the distribution of depreciation charges, there is danger of the accountant deceiving himself into thinking that he has reduced costs when he has merely changed his system of charging depreciation to make the unit cost appear less. If unit cost is to be made a minimum in any useful sense, it is to be done by performing some operation on the machine itself, rather than by adjusting depreciation charges, or by any other process of bookkeeping. The criticism may be answered on Dr. Taylor's behalf by the fact that he uses the test (b) above only to determine the time for the tangible operation of scrapping the machine, and then uses (a) to distribute the depreciation charge over the useful life.

But even if we admit the validity of (a) there remains a question concerning (b). Does the manufacturer desire to make his unit cost plus, in the sense defined, a minimum? Or may not considerations of profit lead him to scrap the machine at some different time from that which makes unit cost plus a minimum, or to shut down his factory or restrict the output of the machine? We shall indeed give a proof that in certain cases the tests (a) and (b) give correct results.

To one important element in the problem I find no allusion in the literature. The value of an old machine or other property must depend upon the operating cost. In all depreciation theories based on unit cost it is assumed that the operating cost is known, and the value is then calculated. But operating cost always includes elements which depend on the value. Hence we cannot know the operating cost until we know the value; that is, until the problem is solved. We shall resolve this difficulty by solving a simple integral equation.

The viewpoint of the present treatment is that the owner wishes to maximize the present value of the output minus the operating costs of the machine or other property. This quantity is, in fact, the value of the property. The first step is, therefore, to set up an expression for the value of the property in terms of value of output, operating costs, and the life of the property. Various hypotheses concerning the economic situation are then applied to evaluate the unknowns appearing in the equation. Depreciation is defined simply as rate of decrease of value. A mathematical formulation of the problem of depreciation is presented from which, it is believed, the whole of the valid portions of existing depreciation theories are readily deducible, and which also lays a foundation for further developments. This treatment is adapted to the consideration of depreciation in connection with variation of such factors as prices and interest rates. The theory is in fact completely general. The property involved will be referred to as a machine purely for convenience of language. The amount of computation involved in a particular application will depend upon the degree of refinement of the data; in the present stage of knowledge of mortality tables of property the work can be made quite simple. A specimen problem is worked out numerically.

Continuous functions, instantaneously compounded interest, and integration will be used as better adapted to the needs of theory than the discontinuous functions and summation processes usually employed. The methods of the integral calculus also appear to me simpler than those of algebraic summation even where only integral numbers of years are involved; while the occurrence of portions of years, which does not seriously affect the integration process, is always a source of trouble when discontinuous functions are used.

Obsolescence is a risk of essentially the same nature as fire, earthquake, or burglary, and should be provided for in the same manner—namely, by an allowance out of operating expenses similar to an insurance premium. Even where no insurance or inadequate insurance is carried, an allowance should be made. Obsolescence is thus to be relegated, with insurance, to the category of operating expenses.

Fundamental Formula

If a machine with operating cost per year $O(\tau)$ produces $Y(\tau)$ units of output per year at time τ and if the value ("theoretical selling price") of a unit of output is x, the annual rental value of the machine at time τ is

$$R(\tau) = x \cdot Y(\tau) - O(\tau).$$

Now the value of a machine is the sum of the anticipated rentals which it will yield, each multiplied by a *discount factor* to allow for interest, plus the scrap or salvage value, also discounted.[3] In the first general case the rate of interest will vary with the time. If $S(n)$ is a function giving the scrap or salvage value at the time n at which the machine is to be discarded, the value at time t is given by the following fundamental formula:

$$V(t) = \int_t^n [xY(\tau) - O(\tau)]e^{-\int_t^\tau \delta(v)\,dv}\,d\tau + S(n)e^{-\int_t^n \delta(v)\,dv}, \tag{1}$$

τ and v being variables of integration representing time. In case the interest

---

[3] The "force of interest" $\delta(t)$ is defined as the rate of increase of an invested sum s divied by s: $\delta(t) = 1/s\ ds/dt$. It follows by integration that $s = s_0 e^{\int_0^t \delta(v)\,dv}$, where s_0 is the value of the invested amount when $t = 0$ and v is a mere variable of integration. Hence $s_0 = se^{-\int_0^t \delta(v)\,dv}$. That is, the present value s_0 of a payment to be made t years hence is the amount s of the expected payment multiplied by the *discount factor* $e^{-\int_0^t \delta(v)\,dv}$. If $\delta(v) = \delta$, a constant, the discount factor is $e^{-t\delta}$. The discount factor takes the form v^t if we write $v = e^{-\delta}$; and $v = 1/(1 + i)$ in this case, where i is the *rate* of interest in the ordinary sense, i.e., the interest payable at the end of a year on each dollar invested for the year. Then

$$\delta = \log_e(1 + i) = i - \frac{i^2}{2} + \frac{i^3}{3} - \cdots$$

and

$$i = e^\delta - 1 = \delta + \frac{\delta^2}{2!} + \frac{\delta^3}{3!} + \cdots.$$

Since both these series converge rapidly they afford a ready means of passing from force of interest to rate of interest or vice versa. The difference between i and δ is ordinarily very small.

The value at time t of a payment to be made at time τ is evidently the amount of the payment multiplied by $e^{-\int_t^\tau \delta(v)\,dv}$. If s is constant, this discount factor equals $v^{\tau-t}$.

rate is constant (1) becomes

$$V(t) = \int_t^n [xY(\tau) - O(\tau)]v^{\tau-t}\, d\tau + S(n)v^{n-t}. \tag{1a}$$

Since the value of a new machine is its cost c, we have for $t = 0$,

$$c = V(0) = \int_0^n [xY(\tau) - O(\tau)]e^{-\int_0^\tau \delta(v)\, dv}\, d\tau + S(n)e^{-\int_0^n \delta(v)\, dv}. \tag{2}$$

Having once bought the machine, the owner wishes to conserve its value as far as he can. This, in the light of (1), is the same as saying that he hopes to get as much net rental as possible out of the machine, interest and scrap value being considered. We, therefore, adopt as the basis of our further work the simple postulate:

I. *Everything in the owner's power will be done to make $V(t)$ a maximum when $t > 0$.*

A second postulate of less general validity, which is often assumed tacitly or is thought to follow from I, is the following:

II. *The machine is always operated at full capacity.*

We shall assume II tentatively.

We suppose that c and the functions Y, O, S and δ of the time are known; but before we can evaluate the right member of (1) we must make some hypothesis which will enable us to determine n and x. There are several possibilities.

(A) If the property has a known useful life n, (2) is a simple equation in x alone.

(B) If x is known, the most obvious way to find a value for n is to solve (2). This, however, can be justified only in very special circumstances. For if x is known in advance of the solution of our problem of depreciation, it must be determined by competitive conditions entirely beyond the control of the machine's owner. He may, for example, use the machine to do work usually done by some other process. In such a case (2) does not hold unless ideally fluid conditions of competition have acted to bring the price c of the machine to such a point as to justify its purchase when n is determined to best advantage. Thus to use (2) to find the best life to allow the machine, the owner must rest precariously upon the judgment of his competition.

The correct procedure is as follows. Let m be the time, if it exists (if not, put ∞ for m), at which the output of the machine ceases to be worth the operating expenses. Then m is determined by the equation $xY(m) - O(m) = 0$. We may for the sake of generality consider that x is a given function $x(t)$ of the time. In this case m is determined by $x(m)\cdot Y(m) - O(m) = 0$, which we may also write $R(m) = 0$. The sum of the possible future rentals of the machine, discounted at (say) a constant rate of interest is then $\int_t^m R(\tau)v^{\tau-t}\, d\tau$. This quantity will decrease until, at time n, it reaches the scrap value $S(n)$. Thus n is determined by $\int_n^m R(\tau)v^{\tau-t}\,d\tau = S(n)$, where $R(m) = 0$. It is of course possible that these equations may have a plurality of solutions. The correct solution is then the one which makes $V(t)$ greatest.

(C) If, as is most common, both x and n are to be determined, we use the equation (2) connecting these unknowns, solving it simultaneously with an additional equation obtained by means of I.

If all economic conditions are static, x is a constant. If economic conditions are not static but their trend can be estimated, we may consider that the value of a unit of output varies in a specified manner with the time, but that its general level depends upon an unknown parameter α. For example, it may be supposed that x increases in geometric progression with known ratio r, but that the initial and final values are unknown. We then write $x = \alpha r^\tau$, or in general, $x = x(\tau, \alpha)$.

Take first the static case. Supposing all our functions continuous and with continuous first derivatives, we differentiate (1) and put $dV(t)/dn = 0$. The result of this operation becomes, after cancelling out the common factor $e^{-\int_t^n \delta(v)\,dv}$ and solving[4] for x,

$$x = \frac{O(n) + \delta(n)S(n) - S'(n)}{Y(n)}, \tag{3}$$

---

[4] The derivation of (3) given in the text is strictly accurate only in the case in which x is independent of n. If, as is generally the case, x is a function of n, the result of equating to zero the derivative of the right member of (1) with regard to n is

$$\frac{dx}{dn}\int_t^n Y(\tau)e^{-\int_t^\tau \delta(v)\,dv}\,d\tau + [xY(n) - O(n) - \delta(n)S(n) + S'(n)]e^{-\int_t^n \delta(v)\,dv} = 0.$$

To get rid of dx/dn we differentiate also the relation (2) between x and n, obtaining an equation exactly like the one above excepting that t is replaced throughout by 0. Eliminating dx/dn between the two we have

$$\begin{vmatrix} \int_t^n Y(\tau)e^{-\int_t^\tau \delta(v)\,dv}\,d\tau & e^{-\int_t^n \delta(v)\,dv} \\ \int_0^n Y(\tau)e^{-\int_0^\tau \delta(v)\,dv}\,d\tau & e^{-\int_0^n \delta(v)\,dv} \end{vmatrix}[xY(n) - O(n) - \delta(n)S(n) + S'(n)] = 0.$$

If the factor in the square brackets vanishes we have (3). But it must vanish, for the determinant disappears, when the upper row is multiplied by $e^{-\int_0^t \delta(v)\,dv}$,

$$\begin{vmatrix} \int_t^n Y(\tau)e^{-\int_0^\tau \delta(v)\,dv}\,d\tau & e^{-\int_0^n \delta(v)\,dv} \\ \int_0^n Y(\tau)e^{-\int_0^\tau \delta(v)\,dv}\,d\tau & e^{-\int_0^n \delta(v)\,dv} \end{vmatrix},$$

which can vanish only if the terms in the first column are equal. This would imply that $\int_0^t Y(\tau)e^{-\int_0^\tau \delta(v)\,dv}\,d\tau = 0$, which is impossible (excepting for $t = 0$) because we suppose $Y(\tau)$ positive for all values of τ less than n.

a condition independent of t, as it should be. This equation states that x, the cost of a unit of product is found by adding the operating cost $O(n)$ of the machine (at the time n when it is least efficient and is about to be scrapped) to interest $\delta(n)S(n)$ on the scrap value and the rate of depreciation $-S'(n)$ of the scrap value, and dividing this sum by the machine's rate of production.

If conditions are not static we may still be able to write a known function $x(\tau, \alpha)$ for x in (1) and (2), and $x(n, \alpha)$ for x in (3). We then solve (2) and (3) simultaneously for n and α and substitute in (1). This parallels a method suggested by Dr. Taylor in the paper cited above, but is slightly more general.

There is of course nothing in the theory to preclude the possibility that n may become infinite.

OPERATING EXPENSE DEPENDENT ON VALUE

It has been assumed up to this point that when we set out to solve a particular depreciation problem we know a function $O(t)$ giving the operating expense at all times in the life of the article. But certain elements of operating cost always depend upon the value of the article at the time, which value we do not know until we have solved the problem. Thus taxes are supposed to be proportional to the value, and insurable risk, if not insurance, certainly is. The risk of obsolescence enters here.

If we write $O(\tau)$ in (1) as a function of $V(\tau)$ and τ we have an *integral equation* to solve for the unknown function $V(t)$. Now the study of integral equations is a new and incomplete branch of mathematics, so that an integral equation written down at random can probably not be solved until pure mathematics has advanced further than at present. By rare good luck, however, the cases which usually arise in practice are of a type leading to one of the few integral equations whose theories are well developed, the *Volterra* equation. This is because the dependence of operating cost upon value is ordinarily *linear*. Thus taxes and insurance premiums are directly proportional to the value, and not to its square or some other function. Hence we may write

$$O(\tau) = A(\tau) + B(\tau)V(\tau),$$

where $A(\tau)$ and $B(\tau)$ are functions which, like $Y(\tau)$ and $\delta(\tau)$, are supposed to have been determined, or at least estimated, on the basis of experience. Substituting this value of $O(\tau)$ in (1) we have

$$V(t) = \int_t^n [x \cdot Y(\tau) - A(\tau)]e^{-\int_t^\tau \delta(v)\,dv}\,d\tau + S(n)e^{-\int_t^n \delta(v)\,dv}$$

$$- \int_t^n B(\tau)V(\tau)e^{-\int_t^\tau \delta(v)\,dv}\,d\tau. \tag{4}$$

This integral equation admits of a very easy solution by reduction to a differential equation as follows. Differentiate (4) and add each member of the resulting equation to the product of $-\delta(t)$ by the corresponding member of (4). Several terms cancel out, leaving[5]

$$\frac{dV(t)}{dt} - [\delta(t) + B(t)]V(t) = -xY(t) + A(t). \tag{5}$$

Let us write for brevity $\gamma(t) = \delta(t) + B(t)$. The solution of (5), found by elementary methods, is then

$$V(t) = e^{\int_0^t \gamma(v)\,dv}\left\{\int_t^n [xY(\tau) - A(\tau)]e^{-\int_0^\tau \gamma(v)\,dv}\,d\tau + k\right\},$$

where k is the constant of integration. To evaluate k we may use either the fact that $V(0) = c$ or that $V(n) = S(n)$. In the first case, we find

$$V(t) = e^{\int_0^t \gamma(v)\,dv}\left\{c - \int_0^t [xY(\tau) - A(\tau)]e^{-\int_0^\tau \gamma(v)\,dv}\,d\tau\right\};$$

in the second case we obtain the equivalent form

$$V(t) = \int_t^n [xY(\tau) - A(\tau)]e^{-\int_t^\tau \gamma(v)\,dv}\,d\tau + S(n)e^{-\int_t^n \gamma(v)\,\delta v}, \tag{1a}$$

which differs from (1) only in that $\gamma(v)$ replaces $\delta(v)$ and $A(\tau)$ replaces $O(\tau)$.

The problem of value and depreciation is now solved exactly as in the simpler case. Equations similar to (2) and (3) are derived and used in the same manner as before, δ being everywhere replaced by γ and $O(t)$ by $A(t)$.

EXAMPLE

Suppose that the data available from experience justify the expectation that

$$Y(\tau) = ae^{-\lambda\tau}, \quad A(\tau) = be^{\mu\tau}, \quad S(n) = 0, \quad \text{and} \quad \delta(v) + B(v) = \gamma, \quad \text{a constant.}$$

Then, by (1a),

$$V(t) = \int_t^n [xae^{-\lambda\tau} - be^{\mu\tau}]e^{-\gamma(\tau-t)}\,d\tau$$

$$= e^{\gamma t}\int_t^n [xae^{-(\lambda+\gamma)\tau} - be^{(\mu-\gamma)\tau}]\,d\tau$$

$$= xa\frac{e^{-\lambda t} - e^{\gamma t - (\lambda+\gamma)n}}{\lambda + \gamma} + b\frac{e^{\mu t} - e^{\gamma t + (\mu-\gamma)n}}{\mu - \gamma}.$$

From an equation similar to (3) we have, since $S(n) = 0$ and therefore

$$S'(n) = 0, \qquad x = \frac{A(n)}{Y(n)} = \frac{be^{(\lambda+\mu)n}}{a}.$$

---

[5] If we solve (5) for x we have the well-known expression for cost of output in terms of operating costs, yield, interest and depreciation on the machine.

Substituting above we have

$$V(t) = \frac{b}{(\lambda + \gamma)(\mu - \gamma)} \left\{ (\mu - \gamma)e^{-\lambda t + (\lambda + \mu)n} - (\lambda + \mu)e^{\gamma t + (\mu - \gamma)n} + (\lambda + \gamma)e^{\mu t} \right\}.$$

For $t = 0$ this gives

$$c = V(0) = \frac{b}{(\lambda + \gamma)(\mu - \gamma)} \left\{ (\mu - \gamma)e^{(\lambda + \mu)n} - (\lambda + \mu)e^{(\mu - \gamma)n} + \lambda + \gamma \right\},$$

which on rearrangement becomes

$$e^{(\lambda + \mu)n} + \frac{\lambda + \mu}{\gamma - \mu}e^{(\mu - \gamma)n} + \frac{\lambda + \gamma}{\mu - \gamma} - (\lambda + \gamma)\frac{c}{b} = 0.$$

This equation for determining n, it will be observed, contains b and c only in the ratio c/b, and does not contain a at all. Since commensurable values may always be taken for γ, λ and μ it reduces to an algebraic equation. It may be proved mathematically that in every case there is just one root for which $n > 0$, and that n approaches zero and infinity with c/b.

Let us take $\gamma = .09$, $\lambda = .06$, $\mu = .14$, $c/b = 20$. The equation becomes $e^{.20n} - 4e^{.05n} = 0$. Hence $e^{.05n} = \sqrt[3]{4}$, and $n = \frac{1}{3} \log 4/.05 \log e = 9.244$ years. Therefore

$$V(t) = \frac{b}{.15}[e^{-.06t + .20n} - 4e^{.09t + .05n} + 3e^{.14t}]$$

$$= \frac{b}{.15}[4^{4/3}(e^{-.06t} - e^{.09t}) + 3e^{.14t}].$$

Calculation from this formula gives the following results for $b = 1,000$.

t	$V(t)$	$V(t-1) - V(t)$
0	20,000.00	
1	16,553.86	3,446.14
2	13,327.50	3,226.36
3	10,345.07	2,982.43
4	7,638.10	2,706.97
5	5,246.63	2,391.57
6	3,220.65	2,025.98
7	1,621.75	1,598.90
8	524.96	1,096.79
9	20.97	503.99
9.244	0	

The decrease in value shown by this table for the first year is more than $1\frac{1}{2}$ times that indicated by the straight line law, and is more than double that calculated by means of the equal annual payment or sinking fund method with interest at 6 per cent.

Depreciation has been defined as rate of decrease of value:

$$D(t) = -\frac{dV(t)}{dt}.$$

The *total depreciation* over a period is the difference between the value at the beginning of the period and that at its end. It equals the average value of the depreciation times the length of the period; for

$$\int_a^b D(t)\,dt = -\int_a^b \frac{dV(t)}{dt}\,dt = V(a) - V(b).$$

The total depreciation in the value of an article over its whole life, is thus $c - S(n)$. The average depreciation is therefore

$$\frac{c - S(n)}{n}. \tag{6}$$

RELATION TO OTHER DEPRECIATION THEORIES

The formula (6) for average depreciation is so simple that if accuracy is not worth while, and if the useful life n is known with some degree of definiteness, it may be assumed that the depreciation of each year is equal to this average depreciation. Or if we are considering the composite depreciation of a large number of similar machines whose times of installation have been uniformly distributed over n years, (6) gives a rough approximation. This *straight line law* is often used in cases in which it gives rise to large errors.

Let us suppose that $O(t)$ and $Y(t)$ are constants q and y, respectively, for a certain number of years of the machine's life and then change abruptly in such a way as to make it clear that the end has come. The supposition that $O(t)$ is constant means, of course, that $B(t) = 0$, that is, that neither risk nor a tax on value exists. Let the scrap value be s, a constant, and let $\delta(t) = \delta$, a constant. Let x also be constant. Then (3) does not hold because its derivation assumed $O(t)$ and $Y(t)$ continuous. But n is now known, so that we can find x from (2) alone.

In this case (1) becomes

$$V(t) = \int_t^n (xy - q)v^{\tau-t}\,d\tau + sv^{n-t}$$

$$= (xy - q)\frac{1 - v^{n-t}}{\delta} + sv^{n-t}.$$

Putting $t = 0$ we have the equation, like (2),

$$c = (xy - q)\frac{1 - v^{-n}}{\delta} + sv^n.$$

The result of eliminating $xy - q$ between these two equations is

$$\begin{vmatrix} V(t) - sv^{n-t} & \dfrac{1 - v^{n-t}}{\delta} \\[2em] c - sv^n & \dfrac{1 - v^n}{\delta} \end{vmatrix} = 0.$$

Now add $-s\delta$ times the second column to the first. Then subtract the second row from the first. Finally, multiplying the second column by $(1 + i)^n$ and remembering that

$$\bar{s}_{\overline{n}|} = \frac{(1 + i)^n - 1}{\delta},$$

we have

$$\begin{vmatrix} V(t) - c & -\bar{s}_{\overline{t}|} \\[1em] c - s & -\bar{s}_{\overline{n}|} \end{vmatrix} = 0.$$

From this we find at once

$$V(t) = c - (c - s)\frac{\bar{s}_{\overline{t}|}}{\bar{s}_{\overline{n}|}} = c - (c - s)\frac{s_{\overline{t}|}}{s_{\overline{n}|}}.$$

Hence

$$V(t) - V(t + 1) = (c - s)\frac{(1 + i)^t}{s_{\overline{n}|}}. \tag{7}$$

This gives a proof of the correctness, under the circumstances assumed, of the "equal annual payment" and "sinking fund" methods of calculating depreciation. But these methods are almost always misleading because the assumptions take no account of the almost universal tendency for operating costs to increase and for output to decrease with age. Consequently they understate the depreciation in the early years.

We may further inquire in what circumstances particular depreciation methods are valid. The test here is supplied by (5), which we now write in the form

$$P(t) = \gamma V(t) - \frac{dV(t)}{dt}, \tag{8}$$

where $P(t) = xY(t) - A(t)$. Let us, for simplicity, consider γ constant, and let us call $P(t)$ the *gross rental*.

According to the straight line method,

$$V(t) = c - (c - s)t/n.$$

Substituting this expression in (8) we find

$$P(t) = c - s + c\gamma - (c - s)\gamma t/n.$$

This is such a special condition that we must in general discard the straight line method as a valid rule where accuracy is required.

The reducing balance method supposes that depreciation is a constant percentage of value. It follows that $V(t) = ce^{-kt}$, k being determined by $s = ce^{-kn}$. From (8) we have in this case the condition $P(t) = ce^{-kt}(\gamma - k)$.

Similarly the condition for the validity of the sinking fund method (7) is

$$P(t) = \gamma - \frac{c - s}{(1 + i)^n - 1}[B(t)(1 + i)^t + \gamma],$$

which is constant if $B(t) = 0$.

Thus the three methods in most common use depend for their validity upon the satisfaction of conditions which involve no disposable constants and which are so special that the chances are overwhelmingly against the satisfaction of any of them in a particular case.

Dr. Taylor[6] defines unit cost plus by a formula which, in terms of continuous functions, becomes

$$x = \frac{O(t) + \delta(t)V(t) - \dfrac{dV(t)}{dt}}{Y(t)},$$

which is equivalent to (5), and reduces to (3) when $t = n$. We may now derive his criterion of minimum unit cost plus as follows. The value of the machine being given by (1) in terms of x and n, and these unknowns being connected by (2), we seek the value of n which will make x a minimum. This can be found from (2) alone. For, differentiating with respect to n and putting $dx/dn = 0$, we have (3). The simultaneous solution of (2) and (3) will then yield the same values of x and n as those found under (C).

Thus Dr. Taylor's treatment yields the same results as those found under (C), save for the discrepancies arising from the use of discontinuous functions which suppose all changes to occur by yearly steps. It is better adapted to calculation than the theory of the present paper in many cases in which $B(t) = 0$, that is, in which there is no element of operating expense such as risk, insurance and taxes. But even under this severe restriction it does not seem a logical foundation for a theory of depreciation without a demonstration, such as the one above, that the life to be allowed a machine to make it most valuable to its owner is also that which makes unit cost plus a minimum. This proposition is not obvious, and is false in case of the failure of our postulate II.

Dr. Taylor in the same paper mentions as an alternative to that discussed— through without advocating it—a theory based on the assumption that "unit cost", defined like unit cost plus except that interest is not included, should be made a minimum. This assumption lacks even the justification given for minimizing unit cost plus. Like the straight line law, it has nothing to recommend it but simplicity.

---

[6] J. S. Taylor, *op. cit.*

FURTHER DEVELOPMENTS

What can be said in case of the failure of postulate II—that the property is used to full capacity? There are very important cases, such as those of mines, in which the postulate is not even approximately true. In this connection we must consider as unknowns not only useful life, value, and depreciation but also the functions $Y(t)$ and $A(t)$. The owner, that is, may voluntarily run the machine at less than full capacity, and wishes us to tell him just how fast to let it run in order that his profits may be a maximum. If we do not tell him, he will guess to the best of his ability. The demand function must be known in order to give a solution.

In all such cases the guiding principle is that the right member of (1), representing discounted future profits, is to be made a maximum. Even if the capitalist system is to give way to one in which service and not profit shall be the object, there will still be an integral of anticipated utilities to be made a maximum. Since we must find a function which maximizes an integral we must in many cases use the Calculus of Variations. But the problem here transcends the questions of depreciation and useful life, and belongs to the dawning economic theory based on considerations of maximum and minimum which bears to the older theories the relations which the Hamiltonian dynamics and the thermodynamics of entropy bear to their predecessors.[7]

The question of charging depreciation as a function of output rather than of time has been discussed of late.[8] It is more natural to consider the depreciation of an automobile in terms of miles than of years. Since interest is an element, this is strictly possible only in case the rate of output $Y(t)$ is known for the whole life of the property; but in this case the arguments commonly advanced for the proposed method fail.

However, an answer may be given to a question of some interest which arises in this connection in case operating cost depends on a controllable rate of output and not merely on the time. What is the additional cost of taking a short trip in an automobile? More generally, what is the additional net cost of a slight increment in output of a machine? We suppose that the increased output extends over a time so short that interest for this period is negligible. The operating cost $A(t)$ per unit of time, exclusive of items such as risk and taxes which are proportional to value, may usually be divided into three classes. One class of costs depends only on the age of the machine, one on age and the amount of current use, and one on the total of past use. We may

---

[7] For the solution of a special problem in the new "entropy" economics, see G. C. Evans, "The Dynamics of Monopoly," *American Mathematical Monthly* XXXI, 2, February 1924, pp. 77–83. A thorough working knowledge of the Calculus of Variations is a prerequisite to the development of this type of economic theory—which doubtless explains why it has not developed further.

All hedonistic and eudaemonistic ethical theories, which declare that the total of pleasure or happiness should be made a maximum, really reduce the question of right conduct to a set of problems in the Calculus of Variations and in the more general theory of maxima of functionals.

[8] E. A. Saliers, *Depreciation, Principles and Applications*, Ronald Press, 1922, pp. 172–178.

thus write as an approximation which is probably good enough for all existing data,

$$A(t) = \alpha(t) + a(t)Y(t) + b \int_0^t Y(\mu)\, d\mu.$$

The additional cost of a small temporary increment z to the total output, concentrated in a time dt, consists partly of increased operating cost at this time and partly of depreciation due to subsequent increase in operating cost. The first part is obviously $za(t)$. The second part depends upon the increment of the integral $\int_0^t Y(\mu)\, d\mu$. This increment will be z for every time τ later than the time t at which the increased use takes place. Thus the future rentals of the machine are decreased by an amount bz per unit of time. The decrease in its value is the discounted sum of these decreases,

$$\int_t^n bze^{-\gamma(\tau-t)}\, d\tau = bz\frac{1 - e^{-\gamma(n-t)}}{\gamma},$$

if γ is constant.

Summary

(A) The value of a machine and that of a unit of its output are *interrelated*, each affecting the other. This economic truism must underlie a correct theory of depreciation.

(B) The fundamental formula (1) gives the value of a machine in terms of time, value of output, operating cost, scrap value, useful life and rate of interest. One relation between these quantities, given by (2), is due to the fact that the cost of the machine when new is its value at that time.

(C) Two postulates are introduced, one equivalent to the supposition of completely rational action with a completely selfish motive; the second, to which we attach less permanent importance, excludes from consideration the possibility that the owner many seek to increase his profits by slowing down production.

(D) Methods are given for finding the value function $V(t)$ when we know either the useful life n of the machine or the value x of a unit of output, and also when we know neither n nor x.

(E) The hitherto untouched difficulty of operating expenses proportional to value is resolved by means of an integral equation.

(F) A numerical example is worked out which is believed typical of a large class of cases.

(G) The depreciation methods hitherto used are tested and all excepting that of Dr. Taylor are found in general to give false results.

(H) When postulate II is abandoned and the rate of output considered controllable, the methods of this paper are capable of further elaboration to cover a great deal of economic and even ethical theory.

(I) Depreciation cannot, as has been proposed, be charged as a function of output alone, the omnipresence of interest preventing this. The increase of depreciation with output can however be calculated by means of our formulae, given adequate experience tables.

Food Research Institute   HAROLD HOTELLING
Stanford University

Since this article was written, two papers bearing on the subject have appeared: "Economics and the Calculus of Variations," by G. C. Evans, *Proceedings of the National Academy of Sciences*, Vol. 11, p. 90; and "A Note on the Theory of Depreciation," by J. S. Taylor, *Bulletin of the American Mathematical Society*, Vol. 31, p. 222.

Stability in Competition[1]

After the work of the late Professor F. Y. Edgeworth one may doubt that anything further can be said on the theory of competition among a small number of entrepreneurs. However, one important feature of actual business seems until recently to have escaped scrutiny. This is the fact that of all the purchasers of a commodity, some buy from one seller, some from another, in spite of moderate differences of price. If the purveyor of an article gradually increases his price while his rivals keep theirs fixed, the diminution in volume of his sales will in general take place continuously rather than in the abrupt way which has tacitly been assumed.

A profound difference in the nature of the stability of a competitive situation results from this fact. We shall examine it with the help of some simple mathematics. The form of the solution will serve also to bring out a number of aspects of a competitive situation whose importance warrants more attention than they have received. Among these features, all illustrated by the same simple case, we find (1) the existence of incomes not properly belonging to any of the categories usually discussed, but resulting from the discontinuity in the increase in the number of sellers with the demand; (2) a socially uneconomical system of prices, leading to needless shipment of goods and kindred deviations from optimum activities; (3) an undue tendency for competitors to imitate each other in quality of goods, in location, and in other essential ways.

Piero Sraffa has discussed[2] the neglected fact that a market is commonly subdivided into regions within each of which one seller is in a quasi-monopolistic position. The consequences of this phenomenon are here considered further. In passing we remark that the asymmetry between supply and demand, between buyer and seller, which Professor Sraffa emphasises is due

---

[1] Presented before the American Mathematical Society at New York, April 6, 1928, and subsequently revised.

[2] "The Laws of Returns Under Competitive Conditions," *Economic Journal,* Vol. XXXVI. pp. 535–550, especially pp. 544 ff. (December 1926).

to the condition that the seller sets the price and the buyers the quantities they will buy. This condition in turn results from the large number of the buyers of a particular commodity as compared with the sellers. Where, as in new oil-fields and in agricultural villages, a few buyers set prices at which they will take all that is offered and exert themselves to induce producers to sell, the situation is reversed. If in the following pages the words "buy" and "sell" be everywhere interchanged, the argument remains equally valid, though applicable to a different class of businesses.

Extensive and difficult applications of the Calculus of Variations in economics have recently been made, sometimes to problems of competition among a small number of entrepreneurs.[3] For this and other reasons a re-examination of stability and related questions, using only elementary mathematics, seems timely.

Duopoly, the condition in which there are two competing merchants, was treated by A. Cournot in 1838.[4] His book went apparently without comment or review for forty-five years until Walras produced his *Théorie Mathématique de la Richesse Sociale*, and Bertrand published a caustic review of both works.[5] Bertrand's criticisms were modified and extended by Edgeworth in his treatment of duopoly in the *Giornale degli Economisti* for 1897,[6] in his criticism of Amoroso,[7] and elsewhere. Indeed all writers since Cournot, except Sraffa and Amoroso,[8] seem to hold that even apart from the likelihood of combination there is an essential instability in duopoly. Now it is true that such competition lacks complete stability; but we shall see that in a very general class of cases the independent actions of two competitors not in collusion lead to a type of equilibrium much less fragile than in the examples of Cournot, Edgeworth and Amoroso. The solution which we shall obtain can break down only in case of an express or tacit understanding which converts the supposed competitors into something like a monopoly, or in case of a price war aimed at eliminating one of them altogether.

Cournot's example was of two proprietors of mineral springs equally available to the market and producing, without cost, mineral water of identical

---

[3] For references to the work of C. F. Roos and G. C. Evans on this subject see the paper by Dr. Roos, "A Dynamical Theory of Economics," in the *Journal of Political Economy*, Vol. XXXV. (1927), or that in the *Transactions of the American Mathematical Society*, Vol. XXX. (1928), p. 360. There is also an application of the Calculus of Variations to depreciation by Dr. Roos in the *Bulletin of the American Mathematical Society*, Vol. XXXIV. (1928), p. 218.

[4] *Recherches sur les Principes Mathématiques de la Théorie des Richesses*. Paris (Hachette). Chapter VII. English translation by N. T. Bacon, with introduction and bibliography by Irving Fisher (New York, Macmillan, 1897 and 1927).

[5] *Journal des Savants* (1883), pp. 499–508.

[6] Republished in English in Edgeworth's *Papers Relating to Political Economy* (London, Macmillan, 1925), Vol. I, pp. 116–26.

[7] *Economic Journal*, Vol. XXXII. (1922), pp. 400–7.

[8] *Lezioni di Economia Matematica* (Bologna, Zanichelli, 1921).

quality. The demand is elastic, and the price is determined by the total amount put on the market. If the respective quantities produced are q_1 and q_2 the price p will be given by a function

$$p = f(q_1 + q_2).$$

The profits of the proprietors are respectively

$$\pi_1 = q_1 f(q_1 + q_2)$$

and

$$\pi_2 = q_2 f(q_1 + q_2).$$

The first proprietor adjusts q_1 so that, when q_2 has its current value, his own profit will be as great as possible. This value of q_1 may be obtained by differentiating π_1, putting

$$f(q_1 + q_2) + q_1 f_1(q_1 + q_2) = 0.$$

In like manner the second proprietor adjusts q_2 so that

$$f(q_1 + q_2) + q_2 f_2(q_1 + q_2) = 0.$$

There can be no equilibrium unless these equations are satisfied simultaneously. Together they determine a definite (and equal) pair of values of q_1 and q_2. Cournot showed graphically how, if a different pair of q's should obtain, each competitor in turn would readjust his production so as to approach as a limit the value given by the solution of the simultaneous equations. He concluded that the actual state of affairs will be given by the common solution, and proceeded to generalise to the case of n competitors.

Against this conclusion Bertrand brought an "objection péremptoire". The solution does not represent equilibrium, for either proprietor can by a slight reduction in price take away all his opponent's business and nearly double his own profits. The other will respond with a still lower price. Only by the use of the quantities as independent variables instead of the prices is the fallacy concealed.

Bertrand's objection was amplified by Edgeworth, who maintained that in the more general case of two monopolists controlling commodities having correlated demand, even though not identical, there is no determinate solution. Edgeworth gave a variety of examples, but nowhere took account of the stabilising effect of masses of consumers placed so as to have a natural preference for one seller or the other. In all his illustrations of competition one merchant can take away his rival's entire business by undercutting his price ever so slightly. Thus discontinuities appear, though a discontinuity, like a vacuum, is abhorred by nature. More typical of real situations is the case in which the quantity sold by each merchant is a continuous function of two variables, his own price and his competitor's. Quite commonly a tiny increase in price by one seller will send only a few customers to the other.

I

The feature of actual business to which, like Professor Sraffa, we draw attention, and which does not seem to have been generally taken account of in economic theory, is the existence with reference to each seller of groups of buyers who will deal with him instead of with his competitors in spite of a difference in price. If a seller increases his price too far he will gradually lose business to his rivals, but he does not lose all his trade instantly when he raises his price only a trifle. Many customers will still prefer to trade with him because they live nearer to his store than to the others, or because they have less freight to pay from his warehouse to their own, or because his mode of doing business is more to their liking, or because he sells other articles which they desire, or because he is a relative or a fellow Elk or Baptist, or on account of some difference in service or quality, or for a combination of reasons. Such circles of customers may be said to make every entrepreneur a monopolist within a limited class and region—and there is no monopoly which is not confined to a limited class and region. The difference between the Standard Oil Company in its prime and the little corner grocery is quantitative rather than qualitative. Between the perfect competition and monopoly of theory lie the actual cases.

It is the gradualness in the shifting of customers from one merchant to another as their prices vary independently which is ignored in the examples worked out by Cournot, Amoroso and Edgeworth. The assumption, implicit in their work, that all buyers deal with the cheapest seller leads to a type of instability which disappears when the quantity sold by each is considered as a continuous function of the differences in price. The use of such a continuous function does, to be sure, seem to violate the doctrine that in one market there can at one time be only one price. But this doctrine is only valid when the commodity in question is absolutely standardised in all respects and when the "market" is a point, without length, breadth or thickness. It is, in fact, analogous to the physical principle that at one point in a body there can at one time be only one temperature. This principle does not prevent different temperatures from existing in different parts of a body at the same time. If it were supposed that any temperature difference, however slight, necessitates a sudden transfer of all the heat in the warmer portion of the body to the colder portion—a transfer which by the same principle would immediately be reversed—then we should have a thermal instability somewhat resembling the instability of the cases of duopoly which have been discussed. To take another physical analogy, the earth is often in astronomical calculations considered as a point, and with substantially accurate results. But the precession of the equinoxes becomes explicable only when account is taken of the ellipsoidal bulge of the earth. So in the theory of value a market is usually considered as a point in which only one price can obtain; but for some purposes it is better to consider a market as an extended region.

Consider the following illustration. The buyers of a commodity will be

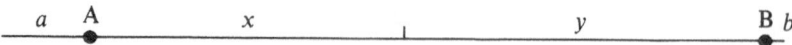

FIGURE 1. Market of length $l = 35$. In this example $a = 4$, $b = 1$, $x = 14$, $y = 16$.

supposed uniformly distributed along a line of length l, which may be Main Street in a town or a transcontinental railroad. At distances a and b respectively from the two ends of this line are the places of business of A and B (Figure 1). Each buyer transports his purchases home at a cost c per unit distance. Without effect upon the generality of our conclusions we shall suppose that the cost of production to A and B is zero, and that unit quantity of the commodity is consumed in each unit of time in each unit of length of line. The demand is thus at the extreme of inelasticity. No customer has any preference for either seller except on the ground of price plus transportation cost. In general there will be many causes leading particular classes of buyers to prefer one seller to another, but the ensemble of such consideration is here symbolised by transportation cost. Denote A's price by p_1, B's by p_2, and let q_1 and q_2 be the respective quantities sold.

Now B's price may be higher than A's, but if B is to sell anything at all he must not let his price exceed A's by more than the cost of transportation from A's place of business to his own. In fact he will keep his price p_2 somewhat below the figure $p_1 - c(l - a - b)$ at which A's goods can be brought to him. Thus he will obtain all the business in the segment of length b at the right of Figure 1, and in addition will sell to all the customers in a segment of length y depending on the difference of prices and lying between himself and A. Likewise A will, if he sells anything, sell to all the buyers in the strips of length a at the left and of length x to the right of A, where x diminishes as $p_1 - p_2$ increases.

The point of division between the regions served by the two entrepreneurs is determined by the condition that at this place it is a matter of indifference whether one buys from A or from B. Equating the delivered prices we have

$$p_1 + cx = p_2 + cy.$$

Another equation between x and y is

$$a + x + y + b = l.$$

Solving we find

$$x = \frac{1}{2}\left(l - a - b + \frac{p_2 - p_1}{c}\right),$$

$$y = \frac{1}{2}\left(l - a - b + \frac{p_1 - p_2}{c}\right),$$

so that the profits are

$$\pi_1 = p_1 q_1 = p_1(a + x) = \frac{1}{2}(l + a - b)p_1 - \frac{p_1^2}{2c} + \frac{p_1 p_2}{2c},$$

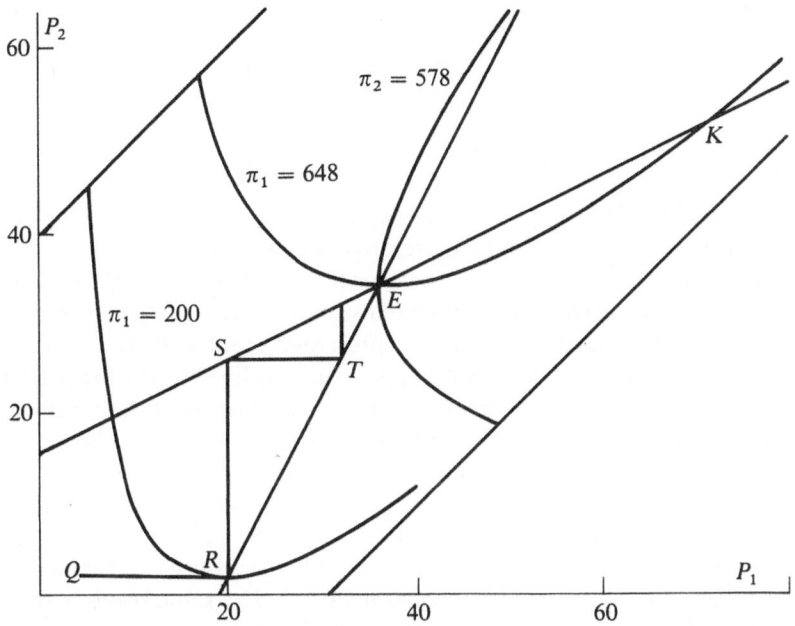

FIGURE 2. Conditions of competition for the market of Figure 1. The co-ordinates represent the prices at A's and B's shops for the same article. The straight lines through E are the two lines of maximum profit. On one of the curves through E, A's profit is everywhere 648; on the other, B's is 578. The lower curve is the locus on which A's profit is 200.

and

$$\pi_2 = p_2 q_2 = p_2(b + y) = \frac{1}{2}(l - a + b)p_2 - \frac{p_2^2}{2c} + \frac{p_1 p_2}{2c}.$$

If p_1 and p_2 be taken as rectangular co-ordinates, each of the last equations represents a family of hyperbolas having identical asymptotes, one hyperbola for each value of π_1 or π_2. Some of these curves are shown in Figure 2, where (as also in Figure 1) we have taken $l = 35$, $a = 4$, $b = 1$, $c = 1$.

Each competitor adjusts his price so that, with the existing value of the other price, his own profit will be a maximum. This gives the equations

$$\frac{\partial \pi_1}{\partial p_1} = \frac{1}{2}(l + a - b) - \frac{p_1}{c} + \frac{p_2}{2c} = 0,$$

$$\frac{\partial \pi_2}{\partial p_2} = \frac{1}{2}(l - a + b) + \frac{p_1}{2c} - \frac{p_2}{c} = 0,$$

from which we obtain

$$p_1 = c\left(l + \frac{a - b}{3}\right),$$

$$p_2 = c\left(l - \frac{a - b}{3}\right);$$

and

$$q_1 = a + x = \frac{1}{2}\left(l + \frac{a - b}{3}\right),$$

$$q_2 = b + y = \frac{1}{2}\left(l - \frac{a - b}{3}\right).$$

The conditions $\partial^2 \pi_1 / \partial p_1^2 < 0$ and $\partial^2 \pi_2 / \partial p_2^2 < 0$, sufficient for a maximum of each of the functions π_1 and π_2, are obviously satisfied.

If the two prices are originally the co-ordinates of the point Q in Figure 2, and if A is the more alert business man of the two, he will change his price so as to make his profit a maximum. This is represented graphically by a horizontal motion to the point R on the line $\partial \pi_1 / \partial p_1 = 0$. This line has the property that every point on it represents a greater profit for A than any other point having the same ordinate. But presently B discovers that his profits can be increased by a vertical motion to the point S on his own line of maximum profit. A now moves horizontally to T. Thus there is a gradual approach to the point E at the intersection of the two lines; its co-ordinates are given by the values of p_1 and p_2 found above. At E there is equilibrium, since neither merchant can now increase his profit by changing his price. The same result is reached if instead of Q the starting point is any on the figure.[9]

Now it is true that prices other than the co-ordinates of the equilibrium point may obtain for a considerable time. Even at this point one merchant may sacrifice his immediate income to raise his price, driving away customers, in the hope that his rival will do likewise and thus increase both profits. Indeed if A moves to the right from E in Figure 2 he may reasonably expect that B will go up to his line of maximum profit. This will make A's profit larger than at E, provided the representing point has not gone so far to the right as K. Without this proviso, A's position will be improved (and so will B's as compared with E) if only B will sufficiently increase p_2. In fact, since the demand is inelastic, we may imagine the two alleged competitors to be amicably exploiting the consumers without limit by raising their prices. The

[9] The solution given above is subject to the limitation that the difference between the prices must not exceed the cost of transportation from A to B. This means that E must lie between the lines $p_1 - p_2 = \pm c(l - a - b)$ on which the hyperbolic arcs shown in Figure 2 terminate. It is easy to find values of the constants for which this condition is not satisfied (for example, $l = 20$, $a = 11$, $b = 8$, $c = 1$). In such a case the equilibrium point will not be E and the expressions for the p's, q's and π's will be different; but there is no essential difference either in the stability of the system or in the essential validity of the subsequent remarks. A's locus of maximum profit no longer coincides with the line $\partial \pi_1 / \partial p_1 = 0$, but consists of the portion of this line above its intersection with $p_1 - p_2 = c(l - a - b)$, and of the latter line below this point. Likewise B's locus of maximum profit consists of the part of the line $\partial \pi_2 / \partial p_2 = 0$ to the right of its intersection with $p_2 - p_1 = c(l - a - b)$, together with the part of the last line to the left of this point. These two loci intersect at the point whose co-ordinates are, for $a > b$,

$$p_1 = c(3l - 3a - b), \qquad p_2 = 2c(l - a),$$

and the type of stability is the same as before.

increases need not be agreed upon in advance but may proceed by alternate steps, each seller in turn making his price higher than the other's, but not high enough to drive away all business. Thus without a formal agreement the rivals may succeed in making themselves virtually a monopoly. Something of a tacit understanding will exist that prices are to be maintained above the level immediately profitable in order to keep profits high in the long run.

But understandings between competitors are notoriously fragile. Let one of these business men, say B, find himself suddenly in need of cash. Immediately at hand he will have a resource: Let him lower his price a little, increasing his sales. His profits will be larger until A decides to stop sacrificing business and lowers his price to the point of maximum profit. B will now be likely to go further in an attempt to recoup, and so the system will descend to the equilibrium position E. Here neither competitor will have any incentive to lower his price further, since the increased business obtainable would fail to compensate him.

Indeed the difficulties of maintaining a price-fixing agreement have often been remarked. Not only may the short-sighted cupidity of one party send the whole system crashing through price-cutting; the very fear of a price cut will bring on a cut. Moreover, a price agreement cannot be made once for all; where conditions of cost or of demand are changing the price needs constant revision. The result is a constant jarring, an always obvious conflict of interests. As a child's pile of blocks falls to its equilibrium position when the table on which it stands is moved, so a movement of economic conditions tends to upset quasi-monopolistic schemes for staying above the point E. For two independent merchants to come to an agreement of any sort is notoriously difficult, but when the agreement must be made all over again at frequent intervals, when each has an incentive for breaking it, and when it is frowned upon by public opinion and must be secret and perhaps illegal, then the pact is not likely to be very durable. The difficulties are, of course, more marked if the competitors are more numerous, but they decidedly are present when there are only two.

The details of the interaction of the prices and sales will, of course, vary widely in different cases. Much will depend upon such market conditions as the degree of secrecy which can be maintained, the degree of possible discrimination among customers, the force of habit and character as affecting the reliance which each competitor feels he can put in the promises of the other, the frequency with which it is feasible to change a price or a rate of production, the relative value to the entrepreneur of immediate and remote profits, and so on. But always there is an insecurity at any point other than the point E which represents equilibrium. Without some agreement, express or tacit, the value of p_1 will be less than or equal to the abscissa of K in Figure 2; and in the absence of a willingness on the part of one of the competitors to forgo immediate profits in order to maintain prices, the prices will become the co-ordinates of E.

One important item should be noticed. The prices may be maintained in a somewhat insecure way *above* their equilibrium values but will never remain *below* them. For if either A or B has a price which is less than that satisfying the simultaneous equations it will pay him *at once* to raise it. This is evident from the figure. Strikingly in contrast with the situation pictured by Bertrand, where prices were for ever being cut below their calculated values, the stabilising effect of the intermediate customers who shift their purchases gradually with changing prices makes itself felt in the existence of a pair of minimum prices. For a prudent investor the difference is all-important.

It is, of course, possible that A, feeling stronger than his opponent and desiring to get rid of him once for all, may reduce his price so far that B will give up the struggle and retire from the business. But during the continuance of this sort of price war A's income will be curtailed more than B's. In any case its possibility does not affect the argument that there is stability, since stability is by definition merely the tendency to return after *small* displacements. A box standing on end is in stable equilibrium, even though it can be tipped over.

II

Having found a solution and acquired some confidence in it, we push the analysis further and draw a number of inferences regarding a competitive situation.

When the values of the p's and q's obtained on pp. 55 and 56 are substituted in the previously found expressions for the profits we have

$$\pi_1 = \frac{c}{2}\left(l + \frac{a-b}{3}\right)^2, \qquad \pi_2 = \frac{c}{2}\left(l - \frac{a-b}{3}\right)^2.$$

The profits as well as the prices depend directly upon c, the unit cost of transportation. These particular merchants would do well, instead of organising improvement clubs and booster associations to better the roads, to make transportation as difficult as possible. Still better would be their situation if they could obtain a protective tariff to hinder the transportation of their commodity between them. Of course, they will not want to impede the transportation of the supplies which come to them; the object of each is merely to attain something approaching a monopoly.

Another observation on the situation is that incomes exist which do not fall strictly within any of the commonly recognised categories. The quantities π_1 and π_2 just determined may be classified as monopoly profits, but only if we are ready to extend the term "monopoly" to include such cases as have been considered, involving the most outright competition for the marginal customer but without discrimination in his favour, and with no sort of open or tacit agreement between the sellers. These profits certainly do not consist of wages, interest or rent, since we have assumed no cost of production. This condition of no cost is not essential to the existence of such profits. If a constant cost of production per unit had been introduced into the calculations above, it would simply have been added to the prices without affecting the profits. Fixed overhead charges are to be subtracted from π_1 and π_2, but may leave a substantial residuum. These gains are not compensation for risk, since they represent a minimum return. They do not belong to the generalised type of "rent", which consists of the advantage of a producer over the marginal producer, since each makes a profit, and since, moreover, we may suppose a and b equal so as to make the situation symmetrical. Indeed π_1 and π_2 represent a special though common sort of profit which results from the fact that the number of sellers is finite. If there are three or more sellers, income of this kind will still exist, but as the number increases it will decline, to be replaced by generalised "rent" for the better-placed producers and poverty for the less fortunate. The number of sellers may be thought of as increasing as a result of a gradual increase in the number of buyers. Profits of the type we have described will exist at all stages of growth excepting those at which a new seller is just entering the field.

As a further problem, suppose that A's location has been fixed but that B is free to choose his place of business. Where will he set up shop? Evidently he will choose b so as to make

$$\pi_2 = \frac{c}{2}\left(l + \frac{b-a}{3}\right)^2$$

as large as possible. This value of b cannot be found by differentiation, as the value thus determined exceeds l and, besides, yields a minimum for π_2 instead of a maximum. But for all smaller values of b, and so for all values of b within the conditions of the problem, π_2 increases with b. Consequently B will seek to make b as large as possible. This means that he will come just as close to A as other conditions permit. Naturally, if A is not exactly in the centre of the line, B will choose the side of A towards the more extensive section of the market, making b greater than a.[10]

This gravitation of B towards A increases B's profit at the expense of A. Indeed, as appears from the expressions on pp. 55 and 56, if b increases so that B approaches A, both q_2 and p_2 increase while q_1 and p_1 diminish. From B's standpoint the sharper competition with A due to proximity is offset by the greater body of buyers with whom he has an advantage. But the danger that the system will be overturned by the elimination of one competitor is

[10] The conclusion that B will tend to gravitate *infinitesimally* close to A requires a slight modification in the particular case before us, but not in general. In the footnote on p. 56 it was seen that when A and B are sufficiently close together, the analytic expressions for the prices, and consequently the profits, are different. By a simple algebraic calculation which will not here be reproduced it is found that B's profits π_2 will increase as B moves from the centre towards A, only if the distance between them is more than four-fifths of the distance from A to the centre. If B approaches more closely his profit is given by $\pi_2 = bc(3l - a - 3b)$, and diminishes with increasing b. This optimum distance from A is, however, an adventitious feature of our problem resulting from a discontinuity which is necessary for simplicity. In general we should consider q_1 and q_2 as continuous functions of p_1 and p_2, instead of supposing, as here, that as $p_2 - p_1$ falls below a certain limit, a great mass of buyers shift suddenly from B to A.

increased. The intermediate segment of the market acts as a cushion as well as a bone of contention; when it disappears we have Cournot's case, and Bertrand's objection applies. Or, returning to the analogy of the box in stable equilibrium though standing on end, the approach of B to A corresponds to a diminution in size of the end of the box.

It has become common for real-estate subdividers in the United States to impose restrictions which tend more or less to fix the character of future businesses in particular locations. Now we find from the calculations above that the total profits of A and B amount to

$$\pi_1 + \pi_2 = c\left[l^2 + \left(\frac{a-b}{3}\right)^2\right].$$

Thus a landlord or realtor who can determine the location of future stores, expecting to absorb their profits in the sales value of the land, has a motive for making the situation as unsymmetrical as possible; for, the more the lack of symmetry, the greater is $(a-b)^2$, which appears in the expression above for $\pi_1 + \pi_2$.

Our example has also an application to the question of capitalism versus socialism, and contributes an argument to the socialist side. Let us consider the efficiency of our pair of merchants in serving the public by calculating the total of transportation charges paid by consumers. These charges for the strip of length a amount to $c\int_0^a t\,dt$, or $\frac{1}{2}ca^2$. Altogether the sum is

$$\tfrac{1}{2}c(a^2 + b^2 + x^2 + y^2).$$

Now if the places of business are both fixed, the quantities a, b and $x + y$ are all determined. The minimum total cost for transportation will be achieved if, for the given value of $x + y$, the expression $x^2 + y^2$ is a minimum. This will be the case if x and y are equal.

But x and y will not be equal unless the prices p_1 and p_2 are equal, and under competition this is not likely to be the case. If we bar the improbable case of A and B having taken up symmetrical positions on the line, the prices which will result from each seeking his own gain have been seen to be different. If the segment a in which A has a clear advantage is greater than b, then A's price will be greater than B's. Consequently some buyers will ship their purchases from B's store, though they are closer to A's, and socially it would be more economical for them to buy from A. If the stores were conducted for public service rather than for profit their prices would be identical in spite of the asymmetry of demand.

If the stores be thought of as movable, the wastefulness of private profit-seeking management becomes even more striking. There are now four variables, a, b, x and y, instead of two. Their sum is the fixed length l, and to minimise the social cost of transportation found above we must make the sum of their squares as small as possible. As before, the variables must be equal. This requires A and B to occupy symmetrical positions at the quartiles of the market. But instead of doing so they crowd together as closely as possible. Even if A, the first in the field, should settle at one of these points, we have seen that B upon his arrival will not go to the other, but will fix upon a location between A and the centre and as near A as possible.[11] Thus some customers will have to transport their goods a distance of more than $\frac{1}{2}l$, whereas with two stores run in the public interest no shipment should be for a greater distance than $\frac{1}{4}l$.

If a third seller C appears, his desire for as large a market as possible will prompt him likewise to take up a position close to A or B, but not between them. By an argument similar to that just used, it may be shown that regard only for the public interest would require A, B and C each to occupy one of the points at distances one-sixth, one-half and five-sixths of the way from one end of the line to the other. As more and more sellers of the same commodity arise, the tendency is not to become distributed in the socially optimum manner but to cluster unduly.

The importance and variety of such agglomerative tendencies become apparent when it is remembered that distance, as we have used it for illustration, is only a figurative term for a great congeries of qualities. Instead of sellers of an identical commodity separated geographically we might have considered two competing cider merchants side by side, one selling a sweeter liquid than the other. If the consumers of cider be thought of as varying by infinitesimal degrees in the sourness they desire we have much the same situation as before. The measure of sourness now replaces distance, while instead of transportation costs there are the degrees of disutility resulting from a consumer getting cider more or less different from what he wants. The foregoing considerations apply, particularly the conclusion that competing sellers tend to become too much alike.

The mathematical analysis thus leads to an observation of wide generality. Buyers are confronted everywhere with an excessive sameness. When a new merchant or manufacturer sets up shop he must not produce something exactly like what is already on the market or he will risk a price war of the type discussed by Bertrand in connection with Cournot's mineral springs. But there is an incentive to make the new product very much like the old, applying some slight change which will seem an improvement to as many buyers as possible without ever going far in this direction. The tremendous standardisation of our furniture, our houses, our clothing, our automobiles and our education are due in part to the economies of large-scale production, in part to fashion and imitation. But over and above these forces is the effect we have been discussing, the tendency to make only slight deviations in order to have for the new commodity as many buyers of the old as possible, to get, so to speak, *between* one's competitors and a mass of customers.

So general is this tendency that it appears in the most diverse fields of

[11] With the unimportant qualification mentioned in the footnote on p. 56.

competitive activity, even quite apart from what is called economic life. In politics it is strikingly exemplified. The competition for votes between the Republican and Democratic parties does not lead to a clear drawing of issues, an adoption of two strongly contrasted positions between which the voter may choose. Instead, each party strives to make its platform as much like the other's as possible. Any radical departure would lose many votes, even though it might lead to stronger commendation of the party by some who would vote for it anyhow. Each candidate "pussyfoots", replies ambiguously to questions, refuses to take a definite stand in any controversy for fear of losing votes. Real differences, if they ever exist, fade gradually with time though the issues may be as important as ever. The Democratic party, once opposed to protective tariffs, moves gradually to a position almost, but not quite, identical with that of the Republicans. It need have no fear of fanatical free-traders, since they will still prefer it to the Republican party, and its advocacy of a continued high tariff will bring it the money and votes of some intermediate groups.

The reasoning, of course, requires modification when applied to the varied conditions of actual life. Our example might have been more complicated. Instead of a uniform distribution of customers along a line we might have assumed a varying density, but with no essential change in conclusions. Instead of a linear market we might suppose the buyers spread out on a plane. Then the customers from one region will patronise A, those from another B. The boundary between the two regions is the locus of points for which the difference of transportation costs from the two shops equals the difference of prices, i.e. for which the delivered price is the same whether the goods are bought from A or from B. If transportation is in straight lines (perhaps by aeroplane) at a cost proportional to the distance, the boundary will be a hyperbola, since a hyperbola is the locus of points such that the difference of distances from the foci is constant. If there are three or more sellers, their regions will be separated from each other by arcs of hyperbolas. If the transportation is not in straight lines, or if its cost is given by such a complicated function as a railroad freight schedule, the boundaries will be of another kind; but we might generalise the term hyperbola (as is done in the differential geometry of curved surfaces) to include these curves also.

The number of dimensions of our picture is increased to three or more when we represent geometrically such characters as sweetness of cider, and instead of transportation costs consider more generally the decrement of utility resulting from the actual commodity being in a different place and condition than the buyer would prefer. Each homogeneous commodity or service or entrepreneur in a competing system can be thought of as a point serving a region separated from other such regions by portions of generalised hyperboloids. The density of demand in this space is in general not uniform, and is restricted to a finite region. It is not necessary that each point representing a service or commodity shall be under the control of a different entrepreneur from every other. On the other hand, everyone who sells an article in different places or who sells different articles in the same place may be said to control the prices at several points of the symbolic space. The mutual gravitation will now take the form of a tendency of the outermost entrepreneurs to approach the cluster.

Two further modifications are important. One arises when it is possible to discriminate among customers, or to sell goods at a delivered price instead of a fixed price at store or factory plus transportation. In such cases, even without an agreement between sellers, a monopoly profit can be collected from some consumers while fierce competition is favouring others. This seems to have been the condition in the cement industry about which a controversy raged a few years ago, and was certainly involved in the railroad rebate scandals.

The other important modification has to do with the elasticity of demand. The problem of the two merchants on a linear market might be varied by supposing that each consumer buys an amount of the commodity in question which depends on the delivered price. If one tries a particular demand function the mathematical complications will now be considerable, but for the most general problems elasticity must be assumed. The difficulty as to whether prices or quantities should be used as independent variables can now be cleared up. This question has troubled many readers of Cournot. The answer is that either set of variables may be used; that the q's may be expressed in terms of the p's, and the p's in terms of the q's. This was not possible in Cournot's example of duopoly, nor heretofore in ours. The sum of our q's was constrained to have the fixed value l, so that they could not be independent, but when the demand is made elastic the constraint vanishes.

With elastic demand the observations we have made on the solution will still for the most part be qualitatively true; but the tendency for B to establish his business excessively close to A will be less marked. The increment in B's sales to his more remote customers when he moves nearer them may be more than compensation to him for abandoning some of his nearer business to A. In this case B will definitely and apart from extraneous circumstances choose a location at some distance from A. But he will not go as far from A as the public welfare would require. The tempting intermediate market will still have an influence.

In the more general problem in which the commodities purveyed differ in many dimensions the situation is the same. The elasticity of demand of particular groups does mitigate the tendency to excessive similarity of competing commodities, but not enough. It leads some factories to make cheap shoes for the poor and others to make expensive shoes for the rich, but all the shoes are too much alike. Our cities become uneconomically large and the business districts within them are too concentrated. Methodist and Presbyterian churches are too much alike; cider is too homogeneous.

Stanford University   HAROLD HOTELLING
California

The Economics of Exhaustible Resources

1. THE PECULIAR PROBLEMS OF MINERAL WEALTH

Contemplation of the world's disappearing supplies of minerals, forests, and other exhaustible assets has led to demands for regulation of their exploitation. The feeling that these products are now too cheap for the good of future generations, that they are being selfishly exploited at too rapid a rate, and that in consequence of their excessive cheapness they are being produced and consumed wastefully has given rise to the conservation movement. The method ordinarily proposed to stop the wholesale devastation of irreplaceable natural resources, or of natural resources replaceable only with difficulty and long delay, is to forbid production at certain times and in certain regions or to hamper production by insisting that obsolete and inefficient methods be continued. The prohibitions against oil and mineral development and cutting timber on certain government lands have this justification, as have also closed seasons for fish and game and statutes forbidding certain highly efficient means of catching fish. Taxation would be a more economic method than publicly ordained inefficiency in the case of purely commercial activities such as mining and fishing for profit, if not also for sport fishing. However, the opposition of those who are making the profits, with the apathy of everyone else, is usually sufficient to prevent the diversion into the public treasury of any considerable part of the proceeds of the exploitation of natural recources.

In contrast to the conservationist belief that a too rapid exploitation of natural resources is taking place, we have the retarding influence of monopolies and combinations, whose growth in industries directly concerned with the exploitation of irreplaceable resources has been striking. If "combinations in restraint of trade" extort high prices from consumers and restrict production, can it be said that their products are too cheap and are being sold too rapidly?

It may seem that the exploitation of an exhaustible natural resource can never be too slow for the public good. For every proposed rate of production there will doubtless be some to point to the ultimate exhaustion which that rate will entail, and to urge more delay. But if it is agreed that the total supply is not to be reserved for our remote descendants and that there is an optimum rate of present production, then the tendency of monopoly and partial monopoly is to keep production below the optimum rate and to exact excessive prices from consumers. The conservation movement, in so far as it aims at absolute prohibitions rather than taxation or regulation in the interest of efficiency, may be accused of playing into the hands of those who are interested in maintaining high prices for the sake of their own pockets rather than of posterity. On the other hand, certain technical conditions most pronounced in the oil industry lead to great wastes of material and to expensive competitive drilling, losses which may be reduced by systems of control which involve delay in production. The government of the United States under the present administration has withdrawn oil lands from entry in order to conserve this asset, and has also taken steps toward prosecuting a group of California oil companies for conspiring to maintain unduly high prices, thus restricting production. Though these moves may at first sight appear contradictory in intent, they are really aimed at two distinct evils, a Scylla and Charybdis between which public policy must be steered.

In addition to these public questions, the economics of exhaustible assets presents a whole forest of intriguing problems. The static-equilibrium type of economic theory which is now so well developed is plainly inadequate for an industry in which the indefinite maintenance of a steady rate of production is a physical impossibility, and which is therefore bound to decline. How much of the proceeds of a mine should be reckoned as income, and how much as return of capital? What is the value of a mine when its contents are supposedly fully known, and what is the effect of uncertainty of estimate? If a mine-owner produces too rapidly, he will depress the price, perhaps to zero. If he produces too slowly, his profits, though larger, may be postponed farther into the future than the rate of interest warrants. Where is his golden mean? And how does this most profitable rate of production vary as exhaustion approaches? Is it more profitable to complete the extraction within a finite time, to extend it indefinitely in such a way that the amount remaining in the mine approaches zero as a limit, or to exploit so slowly that mining operations will not only continue at a diminishing rate forever but leave an amount in the ground which does not approach zero? Suppose the mine is publicly owned. How should exploitation take place for the greatest general good, and how does a course having such an objective compare with that of the profit-seeking entrepreneur? What of the plight of laborers and of subsidiary industries when a mine is exhausted? How can the state, by regulation or taxation, induce the mine-owner to adopt a schedule of production more in harmony with the public good? What about import duties on coal and oil? And for these dynamical systems what becomes of the classic theories of monopoly, duopoly, and free competition?

Problems of exhaustible assets are peculiarly liable to become entangled with the infinite. Not only is there infinite time to consider, but also the possibility that for a necessity the price might increase without limit as the supply vanishes. If we are not to have property of infinite value, we must, in choosing empirical forms for cost and demand curves, take precautions to avoid assumptions, perfectly natural in static problems, which lead to such conditions.

While a complete study of the subject would include semi-replaceable assets such as forests and stocks of fish, ranging gradually downward to such short-time operations as crop carryovers, this paper will be confined in scope to absolutely irreplaceable assets. The forests of a continent occupied by a new population may, for purposes of a first approximation at least, be regarded as composed of two parts, of which one will be replaced after cutting and the other will be consumed without replacement. The first part obeys the laws of static theory; the second, those of the economics of exhaustible assets. Wildlife which may replenish itself if not too rapidly exploited presents questions of a different type.

Problems of exhaustible assets cannot avoid the calculus of variations, including even the most recent researches in this branch of mathematics. However, elementary methods will be sufficient to bring out, in the next few pages, some of the principles of mine economics, with the help of various simplifying assumptions. These will later be generalized in considering a series of cases taking on gradually some of the complexities of the actual situation. We shall assume always that the owner of an exhaustible supply wishes to make the present value of all his future profits a maximum. The force of interest will be denoted by γ, so that $e^{-\gamma t}$ is the present value of a unit of profit to be obtained after time t, interest rates being assumed to remain unchanged in the meantime. The case of variable interest rates gives rise to fairly obvious modifications.[1]

2. FREE COMPETITION

Since it is a matter of indifference to the owner of a mine whether he receives for a unit of his product a price p_0 now or a price $p_0 e^{\gamma t}$ after time t, it is not unreasonable to expect that the price p will be a function of the time of the form $p = p_0 e^{\gamma t}$. This will not apply to monopoly, where the form of the demand function is bound to affect the rate of production, but is characteristic of completely free competition. The various units of the mineral are then to be thought of as being at any time all equally valuable, excepting for varying costs of placing them upon the market. They will be removed and used in order of accessibility, the most cheaply available first. If interest rates or degrees of impatience vary among the mine-owners, this fact will also affect the order of extraction. Here p is to be interpreted as the net price received after paying the cost of extraction and placing upon the market—a convention to which we shall adhere throughout.

---

[1] As in "A General Mathematical Theory of Depreciation," by Harold Hotelling, *Journal of the American Statistical Association*, September, 1925.

The formula

$$p = p_0 e^{\gamma t} \tag{1}$$

fixes the relative prices at different times under free competition. The absolute level, or the value p_0 of the price when $t = 0$, will depend upon demand and upon the total supply of the substance. Denoting the latter by a, and putting

$$q = f(p, t)$$

for the quantity taken at time t if the price is p, we have the equation,

$$\int_0^T q \, dt = \int_0^T f(p_0 e^{\gamma t}, t) \, dt = a, \tag{2}$$

the upper limit T being the time of final exhaustion. Since q will then be zero, we shall have the equation

$$f(p_0 e^{\gamma T}, T) = 0 \tag{3}$$

to determine T.

The nature of these solutions will depend upon the function $f(p, t)$, which gives q. In accordance with the usual assumptions, we shall assume that it is a diminishing function of p, and depends upon the time, if at all, in so simple a fashion that the equations all have unique solutions.

Suppose, for example, that the demand function is given by

$$q = 5 - p \qquad (0 \le p \le 5),$$

$$q = 0 \quad \text{for } p \ge 5,$$

independently of the time.

As q diminishes and approaches zero, p increases toward the value 5, which represents the highest price anyone will pay. Thus at time T,

$$p_0 e^{\gamma T} = 5.$$

The relation (2) between the unknowns p_0 and T becomes in this case

$$a = \int_0^T (5 - p_0 e^{\gamma t}) \, dt = 5T - p_0(e^{\gamma T} - 1)/\gamma.$$

Eliminating p_0, we have

$$a/5 = T + (e^{-\gamma T} - 1)/\gamma,$$

that is,

$$e^{-\gamma T} = 1 + \gamma(a/5 - T).$$

Now, if we plot as functions of T

$$y_1 = e^{-\gamma T}$$

and

$$y_2 = 1 + \gamma(a/5 - T),$$

we have a diminishing exponential curve whose slope where it crosses the y-axis is $-\gamma$, and a straight line with the same slope. The line crosses the y-axis at a higher point than the curve, since, when $T = 0$, $y_1 < y_2$. Hence there is one and only one positive value of T for which $y_1 = y_2$. This value of T gives the time of complete exhaustion. Clearly, it is finite.

If the demand curve is fixed, the question whether the time until exhaustion will be finite or infinite turns upon whether a finite or infinite value of p will be required to make q vanish. For the demand function $q = e^{-bp}$, where b is a constant, the exploitation will continue forever, though of course at a gradually diminishing rate. If $q = \alpha - \beta p$, all will be exhausted in a finite time. In general, the higher the price anticipated when the rate of production becomes extremely small, compared with the price for a more rapid production, the more protracted will be the period of operation.

3. Maximum Social Value and State Interference

As in the static case, there is under free competition in the absence of complicating factors a certain tendency toward maximizing what might be called the "total utility" but is better called the "social value of the resource." For a unit of time this quantity may be defined as

$$u(q) = \int_0^q p(q)\, dq, \tag{4}$$

where the integrand is a diminishing function and the upper limit is the quantity actually placed upon the market and consumed. If future enjoyment be discounted with force of interest γ, the present value is

$$V = \int_0^T u[q(t)]e^{-\gamma t}\, dt.$$

Since $\int_0^T q\, dt$ is fixed, the production schedule $q(t)$ which makes V a maximum must be such that a unit increment in q will increase the integrand as much at one time as at another. That is,

$$\frac{d}{dq} u[q(t)]e^{-\gamma t},$$

which by (4) equals $pe^{-\gamma t}$, is to be a constant. Calling this constant p_0, we have

$$p = p_0 e^{\gamma t},$$

the result (1) obtained in considering free competition. That this gives a genuine maximum appears from the fact that the second derivative is essentially negative, owing to the downward slope of the demand curve.

This conclusion does not, of course, supply any more justification for laissez-faire with the exploitation of natural resources than with other pursuits. It shows that the true basis of the conservation movement is not in any tendency inherent in competition under these ideal conditions. However, there are in extractive industries discrepancies from our assumed conditions leading to particularly wasteful forms of exploitation which might well be regulated in the public interest. We have tacitly assumed all the conditions fully known. Great wastes arise from the suddenness and unexpectedness of mineral discoveries, leading to wild rushes, immensely wasteful socially, to get hold of valuable property.

Of this character is the drilling of "offset wells" along each side of a property line over a newly discovered oil pool. Each owner must drill and get the precious oil quickly, for otherwise his neighbors will get it all. Consequently, great forests of tall derricks rise overnight at a cost of $50,000 or more each; whereas a much smaller number and a slower exploitation would be more economic. Incidentally, great volumes of natural gas and oil are lost because the suddenness of development makes adequate storage impossible.[2]

The unexpectedness of mineral discoveries provides another reason than wastefulness for governmental control and for special taxation. Great profits of a thoroughly adventitious character arise in connection with mineral discoveries, and it is not good public policy to allow such profits to remain in private hands. Of course the prospector may be said to have earned his reward by effort and risk; but can this be said of the landowner who discovers the value of his subsoil purely by observing the results of his neighbors' mining and drilling?

The market rate of interest γ must be used by an entrepreneur in his calculations, but should it be used in determinations of social value and optimum public policy? The use of $\int_0^q p \, dq$ as a measure of social value in a unit of time, whereas the smaller quantity pq would be the greatest possible profit to an owner for the same extraction of material, suggests that a similar integral be used in connection with the various rates of time-preference. There is, however, an important difference between the two cases in that the rate of interest is set by a great variety of forces, chiefly independent of the particular commodity and industry in question, and is not greatly affected by variations in the output of the mine or oil well in question. It is likely, therefore, that in deciding questions of public policy relative to exhaustible resources, no large errors will be made by using the market rate of interest. Of course, changes in this rate are to be anticipated, especially in considering the remote future. If we look ahead to a distant time when all the resources of the earth will be near exhaustion, and the human race reduced to complete poverty, we may expect very high interest rates indeed. But the exhaustion of one or a few types of resources will not bring about this condition.

The discounting of future values of u may be challenged on the ground that future pleasures are ethically equivalent to present pleasure of the same intensity. The reply to this is that capital is productive, that future pleasures are uncertain in a degree increasing with their remoteness in time, and that V

---

[2] Cf. George W. Stoching, *The Oil Industry and the Competitive System*, Hart Schaffner and Marx prize essay (Houghton Mifflin, 1928).

and u are concrete quantities, not symbols for pleasure. They measure the social value of the mine in the sense concerned with the total production of goods, but not properly its utility or the happiness to which it leads, since this depends upon the distribution of wealth, and is greater if the products of the mine benefit chiefly the poor than if they become articles of luxury. A platinum mine is of greater general utility when platinum is used for electrical and chemical purposes than when it is pre-empted by the jewelry trade. However, we must leave questions of distribution of wealth to be dealt with otherwise, perhaps by graded income and inheritance taxes, and consider the effects of various schedules of operation upon the total value of goods produced. It is for this reason that we are concerned with V.

The general question of how much of its income a people should save has been beautifully treated by F. P. Ramsey.[3]

Money metals, of course, occasion very special cause for public concern. Not only does gold production tend to unstabilize prices; but if the uses in the arts can be neglected, the costs of discovery, extraction, and transportation from the mine are, from the social standpoint, wasted.

Still a different reason for caution in deducing a laissez-faire policy from the theoretical maximizing of V under "free" competition is that the actual conditions, even when competition exists, are likely to be far removed from the ideal state we have been postulating. A large producing company can very commonly affect the price by varying its rate of marketing. There is then something of the monopoly element, with a tendency toward undue retardation of production and elevation of price. This will be considered further in our last section. The monopoly problem of course extends also to non-extractive industries; but in dealing with exhaustible resources there are some features of special interest, which will now be examined.

4. Monopoly

The usual theory of monopoly prices deals with the maximum point of the curve

$$y = pq,$$

y being plotted as a function either of p or of q, each of these variables being a diminishing function of the other (Figure 1). We now consider the problem of choosing q as a function of t, subject to the condition

$$\int_0^\infty q \, dt = a, \tag{5}$$

so as to maximize the present value,

$$J = \int_0^\infty qp(q)e^{-\gamma t} \, dt, \tag{6}$$

---

[3] "A Mathematical Theory of Saving," *Economic Journal*, XXXVIII (1928), 543.

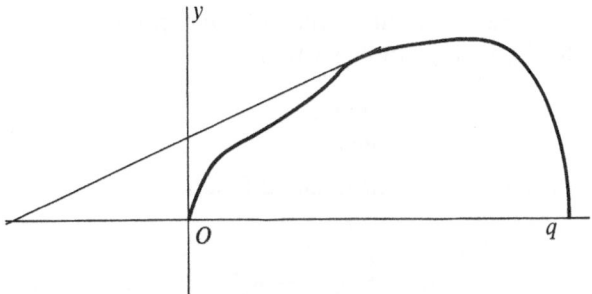

FIGURE 1. $y = pq$. The tangent turns counterclockwise. The value of the mine is proportional to the distance to O from the intersection of the tangent with the q-axis.

of the profits of the owner of a mine. We do not restrict q to be a continuous function of t, though p will be considered a continuous function of q with a continuous first derivative which is nowhere positive. The upper limit of the integrals may be taken as ∞ even if the exploitation is to take place only for a finite time T, for then $q = 0$ when $t > T$.

This may or may not be considered a problem in the calculus of variations; some definitions of that subject would exclude our problem because no derivative is involved under the integral signs, though the methods of the science may be applied to it. However, the problem may be treated fairly simply by observing that

$$qp(q)e^{-\gamma t} - \lambda q, \tag{7}$$

where λ is a Lagrange multiplier, is to be a maximum for every value of t. We must therefore have

$$e^{-\gamma t}\frac{d}{dq}(pq) - \lambda = 0, \tag{8}$$

and also

$$e^{-\gamma t}\frac{d^2}{dq^2}(pq) < 0. \tag{9}$$

Evidently (8) may also be written

$$y' = \frac{d}{dq}(pq) = p + q\frac{dp}{dq} = \lambda e^{\gamma t}, \tag{10}$$

the contrast with the competitive conditions of the last section appearing in the term $q\, dp/dq$.

The constant λ is determined by solving (8) or (10) for q as a function of λ and t and substituting in (5). Upon integrating from 0 to T an equation will then be obtained for λ in terms of T and of the amount a initially in the mine, which is here assumed to be known. The additional equation required to determine T is obtained by putting $q = 0$ for $t = T$.

In general, if p takes on a finite value K as q approaches zero, $q\,dp/dq$ also remaining finite, (8) or (10) can be written

$$\frac{d(pq)}{dq} = Ke^{\gamma(t-T)},$$

Suppose, for example, that the demand function is

$$p = (1 - e^{-Kq})/q$$

$$= K - K^2q/2! + K^3q^2/3! - \ldots,$$

where K is a positive constant. For every positive value of q this expression is positive and has a negative derivative. As q approaches zero, p approaches K. We have

$$y = pq = 1 - e^{-Kq},$$

$$y' = Ke^{-Kq} = \lambda e^{\gamma t},$$

whence

$$q = (\log K/\lambda - \gamma t)/K,$$

this expression holding when t is less than T, the time of ultimate exhaustion. When $t = T$, q is of course zero. We have, therefore, putting $q = 0$ for $t = T$,

$$\log K/\lambda = \gamma T;$$

and from (5)

$$a = \int_0^T (\log K/\lambda - \gamma t)\,dt/K = \gamma \int_0^T (T - t)\,dt/K$$

$$= \gamma T^2/2K$$

so that

$$T = \sqrt{2Ka/\gamma},$$

$$\log K/\lambda = \sqrt{2K\gamma a},$$

giving finally,

$$q = \gamma(\sqrt{2Ka/\gamma} - t)/K.$$

5. GRAPHICAL STUDY: DISCONTINUOUS SOLUTIONS

The interpretation of (10) in terms of Figure 1 is that the rate of production is the abscissa of the point of tangency of a tangent line which rotates counterclockwise. The slope of this line is proportional to a sum increasing at compound interest.

Other graphical representations of the exhaustion of natural resources are possible. Drawing a curve giving $y' = d(pq)/dq$ as a f·· ~· : of q (Figure 2), we have for the most profitable rate of extraction the length of a horizontal line RS which rises like compound interest.

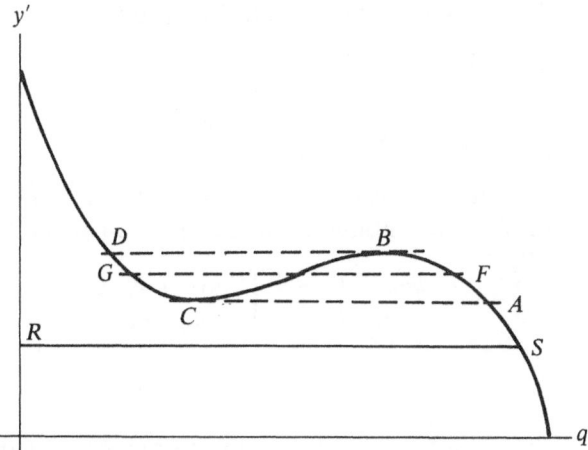

FIGURE 2. *RS* rises with increasing speed. Its length is the rate of production and diminishes.

The waviness with which these curves have been drawn suggests that the solution obtained in this way is not unambiguous. Such waviness will arise if the demand function is, for example

$$p = b - (q - 1)^3, \tag{11}$$

the derivative of which,

$$-3(q - 1)^2,$$

is never positive. Here b is a constant, taken as 1 for Figure 2. For this demand function

$$y' = b - (4q - 1)(q - 1)^2.$$

When the rising line RS reaches the position AC, the point S whose abscissa represents the rate of production might apparently continue along the curve to B and then jump to D; . it might jump from A to C and then move on through D; or it might leave the arc AB at a point between A and B. At first sight there would seem to be another possibility, namely, to jump from A to C, to move up the curve to B, and then to leap to D. But this would mean increasing production for a period. This is never so profitable as to run through the same set of values of q in reverse order, for the total profit would be the same but would be received on the average more quickly if the most rapid production takes place at the beginning of the period. Hence we may regard q as always diminishing, though in this case with a discontinuity.

The values of q between which the leap is made in this case will be determined in Section will be shown that the maximum profit will be reached if the monopolist moves horizontally from a certain point F on AB to a point G on CD.

6. Value of a Mine Monopoly

To find the present value

$$\mathbf{J}_{t_1}^{t_2} = \int_{t_1}^{t_2} pqe^{-\gamma t}\, dt$$

of the profits which are to be realized in any interval t_1 to t_2 during which the maximizing value of q is a continuous function of t, we integrate by parts:

$$\mathbf{J}_{t_1}^{t_2} = -\frac{pqe^{-\gamma t}}{\gamma}\bigg]_{t_1}^{t_2} + \frac{1}{\gamma}\int_{t_1}^{t_2} \frac{d(pq)}{dq}\frac{dq}{dt}e^{-\gamma t}\, dt.$$

When we put

$$y = pq \tag{12}$$

and apply (10), the last integral takes a simple form admitting direct integration. This gives, after applying (10) also to eliminate $e^{-\gamma t}$ from the first term,

$$\mathbf{J}_{t_1}^{t_2} = \frac{\lambda}{\gamma}\left(q - \frac{y}{y'}\right)\bigg]_{t_1}^{t_2}. \tag{13}$$

Now upon differentiating (12), we find

$$qy' = y + q^2 \frac{dp}{dq}.$$

Hence (13) can be written

$$\mathbf{J}_{t_1}^{t_2} = \frac{\lambda}{\gamma}\frac{q^2}{y'}\frac{dp}{dq}\bigg]_{t_1}^{t_2}. \tag{14}$$

The expressions (13) and (14) provide very convenient means of computing the discounted profits. Their validity will be shown in Section 10 to extend to cases in which q is discontinuous.

The expression

$$q - y/y'$$

which appears in (13) is, in terms of Figure 1, the difference between the abscissa and the subtangent of a point on the curve. It therefore equals the distance to the left of the origin of the point where a tangent to the curve meets the q-axis.

The value of the mine when $t = 0$ is, in this notation, $\mathbf{J}_0^T$. It is λ/γ times the distance from the origin to the point of intersection with the negative x-axis of the initial tangent to the curve of monopoly profit.

7. Retardation of Production Under Monopoly

Although the rate of production may suffer discontinuities in spite of the demand function having a continuous derivative, these breaks will always occur during actual production, never at the end. Eventually, q will trail off in a continuous fashion to zero. This means that the highest point of the curve of Figure 2 corresponds to $q = 0$. To prove this, we use the monotonic decreasing character of p as a function of q, which shows that

$$y'(q) = p(q) + qp'(q)$$

is, for positive values of q, less than $p(q)$, and that this in turn is less than $p(0)$. Hence the curve rises higher at the y'-axis than for any level maximum at the right.

The duration of monopolistic exploitation is finite or infinite according as y' takes on a finite or an infinite value when q approaches zero. This condition is a little different from that under competition, where a finite value of p as q approaches zero was found to be necessary and sufficient for a finite time. The two conditions agree unless p remains finite while $qp'(q)$ becomes infinite, in which case the demand curve reaches the p-axis and is tangent to it with contact of order higher than the first. In such a case the period of operation is finite under competition but infinite under monopoly. That this apparently exceptional case is quite likely to exist in fact is indicated by a study of the general properties of supply and demand functions applying the theory of frequency curves, a fascinating subject for which space will not be taken in this paper.

Such a study indicates that very high order contact of the demand curve with the p-axis is to be expected, and therefore that monopolistic exploitation of an exhaustible asset is likely to be protracted immensely longer than competition would bring about or a maximizing of social value would require. This is simply a part of the general tendency for production to be retarded under monopoly.

8. CUMULATED PRODUCTION AFFECTING PRICE

The net price p per unit of product received by the owner of a mine depends not only on the current rate of production but also on past production. The accumulated production affects both cost and demand. The cost of extraction increases as the mine goes deeper; and durable substances, such as gold and diamonds, by their accumulation influence the market. In considering this effect, the calculus of variations cannot be avoided; the following formulation in terms of this science will include as special cases the situations previously treated.

Let x be the amount which has been extracted from a mine, $q = dx/dt$ the current rate of production, and a the amount originally in the mine. Then p is a function of x as well as of q and t. The discounted profit at time $t = 0$, which equals the value of the mine at that time, is

$$\int_0^\infty p(x, q, t)qe^{-\gamma t}\, dt. \tag{15}$$

If exhaustion is to come at a finite time T, we may suppose that $q = 0$ for $t > T$, so that T becomes the upper limit. We put

$$f(x, q, t) = pqe^{-\gamma t}.$$

Then the owner of the mine (who is now assumed to have a monopoly) cannot do better than to adjust his production so that

$$\frac{\partial f}{\partial x} - \frac{d}{dt}\frac{\partial f}{\partial q} = 0.$$

In case f does not involve x, the first term is zero and the former case of monopoly is obtained.

In general the differential equation is of the second order in x, since $q = dx/dt$, and so requires two terminal conditions. One of these is $x = 0$ for $t = 0$. The other end of the curve giving x as a function of t may be anywhere on the line $x = a$, or the curve may have this line as an asymptote. This indefiniteness will be settled by invoking again the condition that the discounted profit is a maximum. The "transversality condition" thus obtained,

$$f - q\frac{\partial f}{\partial q} = 0,$$

that is,

$$q^2 \frac{\partial p}{\partial q} = 0,$$

is equivalent to the proposition that, if p always diminishes when q increases, the curve is tangent or asymptotic to the line $x = a$. Thus ultimately q descends continuously to zero.

Suppose, for example, that q, x, and t all affect the net price *linearly*. Thus

$$p = \alpha - \beta q - cx + gt.$$

Ordinarily α, β, and c will be positive, but g may have either sign. The growth of population and the rising prices to consumers of competing exhaustible goods would lead to a positive value of g. On the other hand, the progress of science might lead to the gradual introduction of new substitutes for the commodity in question, tending to make g negative. The exhaustion of complementary commodities would also tend toward a negative value of g.

The differential equation reduces, for this linear demand function, to the linear form

$$2\beta\frac{d^2x}{dt^2} - 2\beta\gamma\frac{dx}{dt} - c\gamma x = -g\gamma t + g - \alpha\gamma.$$

Since β, c, and γ are positive, the roots of the auxiliary equation are real and of opposite signs. Let m denote the positive and $-n$ the negative root. Since

$$m - n = \gamma,$$

m is numerically greater than n. The solution is

$$x = Ae^{mt} + Be^{-nt} + gt/c - 2\beta g/c^2 - g/c\gamma + \alpha/c,$$

whence

$$q = Ame^{mt} - Bne^{-nt} + g/c.$$

Since $x = 0$ when $t = 0$,

$$A + B - 2\beta g/c^2 - g/c\gamma + \alpha/c = 0.$$

Since $x = a$ and $q = 0$ at the time T of ultimate exhaustion,

$$Ae^{mT} + Be^{-nT} + gT/c - 2\beta g/c^2 - g/c\gamma + \alpha/c - a = 0,$$

$$Ame^{mT} - Bne^{-nT} + g/c \qquad\qquad = 0.$$

From these equations A and B are eliminated by equating to zero the determinant of their coefficients and of the terms not containing A or B. After multiplying the first column by e^{-mT} and the second by e^{nT}, this gives:

$$\Delta = \begin{vmatrix} e^{-mT} & e^{nT} & -2\beta g/c^2 - g/c\gamma + \alpha/c \\ 1 & 1 & gT/c - 2\beta g/c^2 - g/c\gamma + \alpha/c - a \\ m & -n & g/c \end{vmatrix} = 0.$$

Expanding and using the relations $m - n = \gamma$, and $mn = c\gamma/2\beta$, we have for Δ and its derivative with respect to T,

$$\Delta = (e^{-mT} - e^{nT})g/c + (ne^{-mT} + me^{nT})(gT/c - 2\beta g/c^2 - g/c\gamma + \alpha/c - a)$$

$$+ (m + n)(2\beta g/c^2 + g/c\gamma - \alpha/c),$$

$$\Delta' = (e^{nT} - e^{-mT})[T - 1/\gamma + (\alpha - ac)/g]g\gamma/2\beta,$$

the last expression being useful in applying Newton's method to find T. Obviously, the derivative changes sign for only one value of T; for this value Δ has a minimum if g is positive, a maximum if g is negative.

We may measure time in such units that γ, the force of interest, is unity. If money is worth 4 per cent, compounded quarterly the unit of time will then be about 25 years and 1 month. With this convention let us consider an example in which there is an upward secular trend in the price consumers are willing to pay: take $\alpha = 100, \beta = 1, c = 4, g = 16,$ and $a = 10$. The net amount received per unit is in this case

$$p = 100 - q - 4x + 16t.$$

Substituting the values of the constants, and noting that $m = 2$ and $n = 1$, we have

$$\Delta = \begin{vmatrix} e^{-2T} & e^T & 19 \\ 1 & 1 & 4T + 9 \\ 2 & -1 & 4 \end{vmatrix} = (8T + 14)e^T + (4T + 13)e^{-2T} - 57,$$

$$\Delta' = (e^T - e^{-2T})(8T + 22).$$

Evidently $\Delta < 0$ for $T = 0$, $\Delta = +\infty$ for $T = \infty$, and $\Delta' > 0$ for all positive values of T. Hence $\Delta = 0$ has one and only one positive root.

For the trial value $T = 1$ we have

$$\Delta = 5.10, \qquad \Delta' = 77.5.$$

Applying to T the correction $-\Delta/\Delta' = -.07$ roughly, we take $T = .93$ as a second approximation. For this value of T,

$$\Delta = -.06, \qquad \Delta' = 70.0,$$

whence $-\Delta/\Delta' = .001$.

The most profitable schedule of extraction will therefore exhaust the mine in about 0.931 unit of time, or about 23 years and 4 months, perhaps a surprisingly short time in view of the prospect of obtaining an indefinitely higher price in the future, at the rate of increase of 16 per unit of time.

In order that the time of working a mine be infinite, it is necessary not only that the price shall increase indefinitely but that it shall ultimately increase at least as fast as compound interest.

The last two equations for determining A and B now become, since $e^{2T} = 6.4366$ and $e^{-T} = .3942$,

$$6.4366A + .3942B + 12.724 = 0$$

$$12.8732A - .3942B + 4 \qquad = 0.$$

Hence $A = -.866$, $B = -18.13$; so that

$$x = -.866e^{2t} - 18.13e^{-t} + 4t + 19.$$

As a check we observe that this expression for x vanishes when $t = 0$.

Differentiating, we have

$$q = -1.732e^{2t} + 18.13e^{-t} + 4,$$

showing how the rate of production begins at 20.40 and gradually declines to zero. Substitution in the assumed expression for the net price gives

$$p = 100 - q - 4x + 16t$$

$$= 20 + 5.196e^{2t} + 54.39e^{-t},$$

showing a decline from 79.60 at the beginning to 74.90 at exhaustion, owing to the greater cost of extracting the deeper parts of the deposit. The buyer of course pays an increasing, not a decreasing price, namely,

$$p + 4x = 100 - q + 16t$$

$$= 96 + 1.732e^{2t} - 18.13e^{-t} + 16t.$$

This increases from 79.60 to 114.90.

9. THE OPTIMUM COURSE

To examine the course of exploitation of a mine which would be best socially, in contrast with the schedule which a well-informed but entirely selfish owner would adopt, we generalize the considerations of Section 3. Instead of the rate of profit pq, we must now deal with the social return per unit of time,

$$u = \int_0^q p(x, q, t)\, dq,$$

x and t being held constant in the integration. Taking again the market rate of interest as the appropriate discount factor for future enjoyments, we set

$$F = ue^{-\gamma t},$$

and inquire what curve of exploitation will make the total discounted social value,

$$V = \int F\, dt$$

a maximum.

The characteristic equation

$$\frac{\partial F}{\partial x} - \frac{d}{dt}\frac{\partial F}{\partial q} = 0$$

reduces to

$$\frac{\partial p}{\partial q}\frac{d^2 x}{dt^2} + \frac{\partial p}{\partial x}\frac{dx}{dt} - \gamma p = \frac{\partial u}{\partial x} - \frac{\partial p}{\partial t}.$$

The initial condition is $x = 0$ for $t = 0$. The other end-point of the curve is movable on the line $x - a = 0$, a being the amount originally in the mine. The transversality condition,

$$F - q\frac{\partial F}{\partial q} = 0,$$

reduces to

$$u - pq = 0.$$

This is satisfied only for $q = 0$, for otherwise we should have the equation

$$p = \frac{1}{q}\int_0^q p\, dq,$$

stating that the ultimate price is the mean of the potential prices corresponding to lower values of q. Since p is assumed to decrease when q increases, this is impossible. Even if $\partial p/\partial q$ is zero in isolated points, the equation will be impossible if, as is always held, this derivative is elsewhere negative. Hence $q = 0$ at the time of exhaustion.

If, as in Section 8, we suppose the demand function linear,

$$p = \alpha - \beta q - cx + gt,$$

the characteristic equation becomes

$$\beta\frac{d^2 x}{dt^2} - \beta\gamma\frac{dx}{dt} - c\gamma x = -g\gamma t + g - \alpha\gamma.$$

This differs from the corresponding equation for monopoly only in that β is here replaced by $\beta/2$. In a sense, this means that the decline of price, or marginal utility, with increase of supply counts just twice as much in affecting the rate of production, when this is in the control of a monopolist, as the public welfare would warrant.

The analysis of Section 8 may be applied to this case without any qualitative change. The values of m and n depend on β, and are therefore changed. The time T until ultimate exhaustion will be reduced, if social value rather than monopoly profit is to be maximized. For the numerical example given, T was found to be 0.931 unit of time under monopoly. Repeating the calculation for the case in which maximum social value is the goal, we find as the best value only 0.6741 unit of time.

For different values of the constants, even with a linear demand function, the mathematics may be less simple. For example, the equation $\Delta = 0$ may have two positive roots instead of one. This will be the case if the numerical illustration chosen be varied by supposing that the sign of g is reversed, owing to the progressive discovery of substitutes, the direct effect of passage of time being then to decrease instead of increase the price. In such cases a further examination is necessary of the two possible curves of development, to determine which will yield a greater monopoly profit or total discounted social value, according to our object.

10. Discontinuous Solutions

Even if the rate of production q has a discontinuity, as in the example of Section 5, the condition that $\int f\, dt$ shall be a maximum requires that each of the quantities

$$\frac{\partial f}{\partial q}, \qquad f - q\frac{\partial f}{\partial q},$$

must nevertheless be continuous.[4] This will be true whether f stands for discounted monopoly profit or discounted total utility.

The equation (8) on p. 71, may be written

$$\frac{\partial f}{\partial q} = \lambda,$$

which shows, since the left-hand member is continuous, that λ must have the same value before and after the discontinuity.

When p is a function of q alone, the two continuous quantities may be written in the notation of Section 4, $y'e^{-\gamma t}$ and $(y - qy')e^{-\gamma t}$, which shows that y' and $y - qy'$ are continuous. Thus the expression $\lambda(q - y/y')$ appearing in (13), p. 74, is continuous. Consequently the expressions (13) or (14) pertaining to the different time-intervals may simply be added to obtain an expression

---

[4] C. Carathéodory, "Über die diskontinuirlichen Lösungen in der Variationsrechnung," thesis, Göttingen, 1904, p. 11. The condition that the first of these quantities must be continuous is given in the textbooks, but for some reason the second is generally omitted.

of the same form. Hence the present value of the discounted future profits the mine—and therefore of the mine—is in such cases the difference between the values of

$$\lambda(q - y/y')/\gamma$$

at present and at the time of exhaustion.

We are now ready to answer such questions as that raised at the end of Section 5 as to the location of the discontinuity there shown to exist in the most profitable schedule of production when the demand function is

$$p = b - (q - 1)^3.$$

Since in this case

$$f = pqe^{-\gamma t} = [bq - q(q - 1)^3]e^{-\gamma t},$$

the two quantities

$$b - (4q - 1)(q - 1)^2,$$

$$3q^2(q - 1)^2,$$

are continuous. Consequently

$$(4q - 1)(q - 1)^2$$

and

$$q^2(q - 1)^2$$

are continuous. If q_1 denote the rate of production just before the sudden jump and q_2 the initial rate after it, this means that

$$(4q_1 - 1)(q_1 - 1)^2 = (4q_2 - 1)(q_2 - 1)^2,$$

$$q_1^2(q_1 - 1)^2 = q_2^2(q_2 - 1)^2.$$

The only admissible solution is:

$$q_1 = (3 + \sqrt{3})/4 = 1.1830, \qquad q_2 = (3 - \sqrt{3})/4 = 0.31699.$$

11. TESTS FOR A TRUE MAXIMUM

The equations which have been given for finding the production schedule of maximum profit or social value are necessary, not sufficient, conditions for maxima, like the vanishing of the first derivative in the differential calculus. We must also consider more definitive tests.

The integrals which have arisen in the problems of exhaustible assets are to be maxima, not necessarily for the most general type of variation conceivable for a curve, but only for the so-called "special weak" variations. The nature of the economic situation seems to preclude all variations which involve turning time backward, increasing the rate of production, maintaining two different rates of production at the same time, or varying production with infinite rapidity. Extremely sudden increases in production usually involve special costs which will be borne only under unexpected conditions, and are to be avoided in long-term planning. Likewise sudden decreases involve social losses of great magnitude such as unemployment, which even a selfish monopolist will often try to prevent. This will be considered further in the next section. It is indeed possible that in some special cases these "strong" variations might take on some economic significance, but such a situation would involve forces of a different sort from those with which economic theory is ordinarily concerned.

The critical tests which must be applied are by the foregoing considerations reduced to two—those of Legendre and Jacobi.[5] The Legendre test requires, in order that the total discounted utility or social value (Section 9) shall be a maximum, that

$$\frac{\partial^2 u}{\partial q^2} = \frac{\partial p}{\partial q} < 0,$$

a condition which is always held to obtain save in exceptional cases. In order that the chosen curve shall yield a genuine maximum for a monopolist's profit, the Legendre test requires that

$$\frac{\partial^2 (pq)}{\partial q^2} = 2\frac{\partial p}{\partial q} + q\frac{\partial^2 p}{\partial q^2} < 0.$$

This means that the curve of Figure 1 is convex upward at all points touched by the turning tangent. The re-entrant portions, if any, are passed over, producing discontinuities in the rate of production.

When the solution of the characteristic equation has been found in the form

$$x = \varphi(t, A, B),$$

A and B being arbitrary constants, the Jacobi test requires that

$$\frac{\partial \varphi / \partial A}{\partial \varphi / \partial B}$$

shall not take the same value for two different values of t. For the example of Section 8 this critical quantity is simply $e^{(m+n)t}$, which obviously satisfies the test. The solution represents a real, not an illusory maximum for the monopolist's profit. The like is true for the schedule of production maximizing the total discounted utility with the same demand function. Each case must, however, be examined separately, as the test might show in some instances that a seeming maximum could be improved.

12. The Need for Steadiness in Production

The demand function giving p may involve not only the rate of production q, but also the rate of change q' of q. Such a condition would display a duality with that considered by C. F. Roos[6] and G. C. Evans,[7] who hold that the

---

[5] A. R. Forsyth, *Calculus of Variations* (Cambridge, 1927), pp. 17–28.

[6] "A Dynamical Theory of Economics," *Journal of Political Economy*, XXXV (1927), 632, and references there given; also "A Mathematical Theory of Depreciation and Replacement," *American Journal of Mathematics*, L (1928), 147.

[7] "The Dynamics of Monopoly," *American Mathematical Monthly*, Vol. XXXI (1924); also *Mathematical Introduction to Economics* (McGraw-Hill Book Co., 1930).

quantity of a commodity which can be sold per unit of time depends ordinarily upon the rate of change of the price, as well as upon the price itself. If p is a function of x, q, q', and t, the maximum of monopoly profit or of social value can only be obtained if the course of exploitation satisfies a fourth-order differential equation.

More generally we might suppose that p and its rate of change p' are connected with x, q, q', and t by a relation

$$\varphi(p, p', x, q, q', t) = 0.$$

This presents a Lagrange problem, which can be dealt with by known methods.[8] A further generation is to suppose that the price, the quantity, and their derivatives are subject to a relation in the nature of a demand function which also involves an integral or integrals giving the effect of past prices and rates of consumption.[9]

Capital investment in developing the mine and industries essential to it is a source of a need for steady production; the desirability of regular employment for labor is another. Under the term "capital" might possibly be included the costs, both to employers and to laborers, in drawing laborers to the mine from other places and occupations. The returning of these laborers to other occupations as production declines would have to be reckoned as part of the social cost. Whether this would enter into the mine-owner's costs would probably depend upon whether the laborers have at the beginning sufficient information and bargaining power to insist upon compensation for the cost to them of the return shift.

Problems in which the fixity of capital investment plays a part in determining production schedules may be dealt with by introducing new variables $x_1, x_2, \ldots$, to represent the various types of capital investment involved. In so far as these variables are continuous, the problem is that of maximizing an integral involving $x, x_1, x_2, \ldots$, and their derivatives, using well-known methods. The simultaneous equations

$$\frac{\partial f}{\partial x_i} - \frac{d}{dt}\frac{\partial f}{\partial x_i'} = 0 \qquad (i = 0, 1, 2, \ldots; x_0 = x; x_i' = dx_i/dt)$$

are necessary for a maximum. The depreciation of mining equipment raises considerations of this kind.

The cases considered in the earlier part of this paper all led to solutions in which the rate of production of a mine always decreases. By considering the influence of fixed investments and the cost of accelerating production at the beginning, we may be led to production curves which rise continuously from zero to a maximum, and then fall more slowly as exhaustion approaches.

[8] Cf. G. A. Bliss, "The Problem of Lagrange in the Calculus of Variations" (mimeographed by O. E. Brown), (University of Chicago Bookstore).

[9] See "Generalized Lagrange Problems in the Calculus of Variations," by C. F. Roos, *Transactions of the American Mathematical Society*, XXX, (1927), 360.

Certain production curves of this type have been found statistically to exist for whole industries of the extractive type, such as petroleum production.[10]

13. CAPITAL VALUE TAXES AND SEVERANCE TAXES

An unanticipated tax upon the value of a mine will have no effect other than to transfer to the government treasury a part of the mine-owner's income. An anticipated tax at the rate α per year and payable continuously will have the same effects upon the value of the mine and the schedule of production as an increase of the force of interest by α. This we shall now prove.

From the income pq from the mine at time t must now be deducted the tax, $\alpha J(t)$. Consequently the value at time τ is

$$J(\tau) = \int_\tau^T [pq - \alpha J(t)]e^{-\gamma(t-\tau)}\, dt.$$

This integral equation in J reduces by differentiation to a differential equation:

$$J'(\tau) = -pq + \alpha J(\tau) + \gamma J(\tau).$$

The solution is found by well-known methods. The constant of integration is evaluated by means of the condition that $J(T) = 0$. We have:

$$J(\tau) = \int_\tau^T pqe^{-(\alpha+\gamma)(t-\tau)}\, dt,$$

so that α is merely added to γ.

Quite a different kind of levy is represented by the "severance tax."[11] Such a tax, of so much per unit of material extracted from the mine, tends to conservation. The ordinary theory of monopoly of an inexhaustible commodity suggests that the incidence of such a tax is divided between monopolist and consumer, equally in the case of a linear demand function. However, for an exhaustible supply the division is in a different proportion, varying with time and the supply remaining. Indeed, the imposition of the tax will lead eventually to an actually lower price than as if there had been no tax.

[10] C. E. Van Orstrand, "On the Empirical Representation of Certain Production Curves," *Journal of the Washington Academy of Sciences*, XV, (1925), 19.

[11] A variant is an ad valorem tax. A great deal of information and discussion concerning these taxes is contained in the biennial *Report of the Minnesota State Tax Commission*, 1928 (St. Paul). From p. 111 of this report it appears that Alabama since 1927 has had a severance tax of $2\frac{1}{2}$ cents a ton on coal, $4\frac{1}{2}$ cents a ton on iron ore, and 3 per cent on quarry products; Montana taxes coal extracted at 5 cents per ton; Arkansas imposes a tax of $2\frac{1}{2}$ per cent on the gross value of all natural resources except coal and timber, 1 per cent on coal, and 7 cents per 1,000 board feet on timber. Minnesota taxes iron ore extracted at 6 per cent on value minus cost of labor and materials used in mining, and also assesses ore lands at a higher rate than other property for the general property tax. These taxes are not based entirely on the conservation idea, but aim also at taxing persons outside the state, or "retaining for the state its natural heritage." Since Minnesota produces about two-thirds of the iron ore of the United States, the outside incidence is doubtless accomplished. Mexican petroleum taxes have the same object. The Minnesota Commission believes that prospecting for ore has virtually ceased on account of the high taxes.

Consider the linear demand function

$$p = \alpha - \beta q$$

and, for simplicity, no cost of production. The rate of net profit, after paying a tax v per unit extracted, will be

$$(p - v)q = (\alpha - v)q - \beta q^2.$$

As in Section 4, the derivative increases as compound interest:

$$\alpha - v - 2\beta q = \lambda e^{\gamma t}.$$

Since ultimately $q = 0$ and $t = T$, we obtain

$$\alpha - v = \lambda e^{\gamma T},$$

whence eliminating λ and solving for q,

$$q = [1 - e^{\gamma(t-T)}](\alpha - v)/2\beta.$$

The time of exhaustion T is related to the amount originally in the mine through the equation

$$a = \int_0^T q\, dt = (\gamma T + e^{-\gamma T} - 1)(\alpha - v)/2\beta\gamma;$$

whence

$$dT = \frac{2\beta a\, dv}{(\alpha - v)^2(1 - e^{-\gamma T})},$$

showing how much of an increase in time of exploitation is likely to result from the imposition of a small severance tax. The effect upon the rate of production at time t is

$$dq = \frac{\partial q}{\partial v}\, dv + \frac{\partial q}{\partial T}\, dT$$

$$= dv\{-1 + e^{\gamma(t-T)}[1 + 2\beta\gamma a/(\alpha - v)(1 - e^{-\gamma T})]\}/2\beta.$$

From the form of the demand function it follows that the increase in price at time t is

$$dp = -\beta\, dq = dv\{\tfrac{1}{2} - e^{\gamma(t-T)}[\tfrac{1}{2} + \beta\gamma a/(\alpha - v)(1 - e^{-\gamma T})]\}.$$

If a is very large, then so is T; the expression in curly brackets will, for moderate values of t, differ infinitesimally from $\tfrac{1}{2}$, reducing to the case of monopoly with unlimited supplies. However, dp will always be less than $\tfrac{1}{2}\, dv$ and, as exhaustion approaches, will decline and become negative. Finally, when $t = T$, the price of the tax-paid articles to buyers is lower by

$$\beta\gamma av/(\alpha - v)(1 - e^{-\gamma T})$$

than the ultimate price if there had been no tax. The price will, nevertheless, be so high that very little of the commodity will be bought.

A tax on a monopolist which will lead him to reduce his prices is reminiscent of Edgeworth's paradox of a tax on first-class railway tickets which makes the monopolistic (and unregulated) owner's most profitable course the reduction of the prices both of first- and of third-class tickets, besides paying the tax himself.[12] The case of a mine is, however, of a distinct species from Edgeworth's, and cannot be assimilated to it by treating ore extracted at different times as different commodities. Indeed, in the simple case of mine economics which we are now considering, the demands at different times are not correlated; supplies put upon the market now and in the future neither complement nor compete with each other. Correlated demand of a particular type was, on the other hand, an essential feature of Edgeworth's phenomenon.

The ultimate lowering of price and the extension of the life of a mine as a result of a severance tax are not peculiar to the linear demand function, but hold similarly for any declining demand function $p(q)$ whose slope is always finite. This general proposition does not depend upon the tax being small.

The conclusion reached in the linear case that the division of the incidence of the tax is more favorable to the consumer than for an inexhaustible supply is probably true in general; at least this is indicated by an examination of a number of demand curves. However, the general proposition seems very difficult to prove.

Since the severance tax postpones exhaustion, falls in considerable part on the monopolist, and leads ultimately to an actual lowering of price, it would seem to be a good tax. It is particularly to be commended if the monopolist is regarded as *unfairly* possessed of his property, and there is no other feasible means of taking away from him so great a portion of it as the severance tax will yield. However, the total wealth of the community may be diminished rather than increased by such a tax. Considering as in Section 3 the integral u of the prices p which buyers are willing to pay for quantities below that actually put on the market, and the time-integral U of values of u discounted for interest, we have in the case of linear demand just discussed,

$$u = \int_0^q (\alpha - \beta q)\, dq = \alpha q - \frac{1}{2}\beta q^2.$$

If we were considering the portion of this social benefit which inures to consumers, we should have to subtract the portion pq which they pay to the monopolist, an amount from which he would have to subtract the tax, which benefits the state. But the sum of all these benefits is u, which is affected by the tax only as this affects the rate of production q.

If, for simplicity, we measure time in such units that $\gamma = 1$, the rate of production determined earlier in this section becomes

$$q = (1 - e^{t-T})(\alpha - v)/2\beta.$$

---

[12] *Economic Journal*, VII (1897), 231, and various passages in Edgeworth's *Papers Relating to Political Economy*.

Substituting this in the expression for u and the result in U, we obtain

$$U = \int_0^T u e^{-t}\, dt$$

$$= (\alpha - v)[4\alpha(1 - e^{-T} - Te^{-T}) - (\alpha - v)(1 - 2Te^{-T} - e^{-2T})]/8\beta.$$

We differentiate U and then, to examine the effect of a small tax, put $v = 0$. The results simplify to:

$$\frac{\partial U}{\partial v} = -(1 - e^{-T})^2 \alpha/4\beta,$$

$$\frac{\partial U}{\partial T} = [(T + 1)e^{-T} - e^{-2T}]\alpha^2/4\beta.$$

From the amount initially in the mine,

$$a = (T + e^{-T} - 1)/2\beta,$$

we obtain, as on page 85,

$$\frac{dT}{dv} = \frac{2\beta a}{\alpha^2(1 - e^{-T})},$$

when $v = 0$. Substituting here the preceding value for a we find, after simplification,

$$\frac{dU}{dv} = \frac{\partial U}{\partial v} + \frac{\partial U}{\partial T}\frac{dT}{dv} = -\frac{\alpha}{4\beta}\frac{e^T + e^{-T} - 2 - T^2}{e^T - 1}.$$

The numerator of the last fraction may be expanded in a convergent series of powers of T in which all the terms are positive. Hence dU/dv is negative.

Thus a small tax on a monopolized resource will diminish its total social value, at least if the demand function is linear. Whether this is true for demand functions in general is an unsolved problem.

We have here supposed the tax v to be constant, permanent, and fully foreseen. Since an unforeseen tax will have unforeseen results, we can scarcely build up a general theory of such taxes. However, any tax of amount varying with time in a manner definitely fixed upon in advance will have predictable results. In this connection, an interesting problem is to fix upon a schedule of taxation v, which may involve the rate of production q and the cumulated production x, as well as the time, such that, when the monopolist then chooses his schedule of production to maximize his profit, the social value U will be greater than as if any other tax schedule had been adopted. This leads to a problem of Lagrange type in the calculus of variations, one end-point being variable. Putting $q = dx/dt$ and

$$J = \int_0^T f(x, q, v, t)\, dt, \qquad U = \int_0^T F(x, q, t)\, dt,$$

the problem is to choose v, subject to the differential relation

$$\frac{\partial f}{\partial x} + \frac{\partial f}{\partial v}\frac{\partial v}{\partial x} - \frac{d}{dt}\left(\frac{\partial f}{\partial q} + \frac{\partial f}{\partial v}\frac{\partial v}{\partial q}\right) = 0,$$

so that U will be a maximum. In general, of course, a still greater value of U would be obtainable, at least in theory, by public ownership and operation.

14. MINE INCOME AND DEPLETION

With income taxes we are not concerned except for the determination of the amount of the income from a mine. The problem of allowance for depletion has been a perplexing one. It has been said that if the value of ore removed from the ground could be claimed as a deduction from income, then a mining company having no income except from the sale of ore could escape payment of income tax entirely. The fallacy of this contention may be examined by considering the value of the mine at time t,

$$J(t) = \int^{T} pqe^{-\gamma(\tau - t)}\, d\tau.$$

In this integral p and q have the values corresponding to the time τ, later than t, assigned by whatever production schedule has been adopted, whether this results from competition, from a desire to maximize monopoly profit, or from any other set of conditions. The net income consists of the return from sales of material removed (cost of production and selling having as usual been deducted), minus the decrease in the value of the mine. It therefore equals, per unit of time,

$$pq + dJ/dt;$$

and from the expression for J, this is exactly γJ. In other words, any particular production schedule fixes the value of the mine at such a figure that the income at any time, after allowing for depletion, is exactly equal to the interest on the value of the investment at that time.

But, although the rate of decline in value of a mine seems a logical quantity to define as depletion and to deduct from income, such is not the practice of income-tax administrations, at least in the United States. The value of the property upon acquisition, or on 1 March, 1913, a date shortly before the inauguration of the tax, if acquired before that time, is taken as a basis and divided by the number of units of material estimated to have been in the ground at that time. The resulting "unit of depletion," an amount of money, is multiplied by the number of tons, pounds, or ounces of material removed in a year to give the depletion for that year. The total of depletion allowances must not exceed the original value of the property.

The differences between the two methods of calculating depletion arise from the uncertainties of valuation and of forecasting price, demand, production, costs, interest rates, and amount of material remaining. If the theoretical method were applied, a year in which the mine failed to operate would still be set down as yielding an income equal to the interest on the investment value. This seems anomalous only because of another defect, from the theoretical standpoint, in income-tax laws: the non-taxation of increase in value of a property until the sale of the property. During a year of idleness, if foreseen, the value of the property is actually increasing, for the idle year has been considered in fixing the value at the beginning of the year.

An amendment made to the United States federal income-tax law in 1918 provides that the valuation upon which depletion is calculated may under certain circumstances be taken, not as the value of the property when acquired or in 1913, but the higher value which it later took when its mineral content was discovered. This provision has the effect of materially increasing depletion allowances, and so of reducing tax payments. The sudden increase in value when the mineral is discovered might well be regarded as taxable income, but is not so regarded by the law unless the property is immediately sold. The framers of the statute seem indeed, according to its language, to have considered this increase in value a reward for the efforts and risks of prospecting, which would suggest that it is of the nature of income, a reasonable position. However, the object of the amendment is to treat this increment as pre-existing captial value, to be returned to the owner by sale of the mineral. The amendment appears to be inconsistent and quite too generous to the owners particularly affected.

15. DUOPOLY

Intermediate between monopoly and perfect competition, and more closely related than either to the real economic world, is the condition in which there are a few competing sellers. In a former paper[13] this situation was discussed for the static case, with special reference to a factor usually ignored, the existence with reference to each seller of groups of buyers who have a special advantage in dealing with him in spite of possible lower prices elsewhere. More than one price in the same market is then possible, and with a sort of quasi-stability which sets a lower limit to prices, as well as the known upper limit of monopoly price.

For exhaustible resources the corresponding problems of competition among a small number of entrepreneurs may be studied in the first instance by means of the jointly stationary values of the several integrals representing discounted profits. We need not confine ourselves, as we have done for convenience in dealing with monopoly, to a single mine for each competitor. Let there be m competitors, and let the one numbered i control n_i mines, whose production rates and initial contents we shall denote by $q_{i1}, \ldots, q_{in_i}$, and $a_{i1}, \ldots, a_{in_i}$, respectively. The demand functions will be intercorrelated, both among the mines owned by each competitor and between the mines of different concerns. Consequently the m integrals J_i representing the discounted profits

---

[13] Harold Hotelling, "Stability in Competition," *Economic Journal*, XLI (March, 1929), 41.

will involve in their integrands f_i all the q_{ij}, as well as some at least of the cumulated productions $x_{ij} = \int q_{ij}\, dt$. If the ith owner wishes to make his profit a maximum, assuming the production rates of the others to have been fixed upon, he will adjust his n_i production rates so that

$$\frac{\partial f_i}{\partial x_{ij}} - \frac{d}{dt}\frac{\partial f_i}{\partial q_{ij}} = 0 \qquad (j = 1, 2, \ldots, n_i).$$

Continuing the analogy with the static case, we are to imagine that the other competitors, hearing of his plans, do likewise, altering their schedules to conform to equations resembling those above. When the ith owner learns of their changed plans, he will in turn readjust. The only possible final equilibrium with a settled schedule of production for each mine will be determined by the solution of the set of differential equations of this type, which are exactly as numerous as the mines, and therefore as the variables to be determined. All this is a direct generalization of the case of inexhaustible supplies. But we shall show that the solution tends to overstate the production rates and understate the prices of competing mines.

Doubts in plenty have been cast upon the result in the simpler case, and the reasons which can there be adduced in favor of the solution are even more painfully inadequate when the supplies are of limited amount. The chief difficulty with the problem of a small number of sellers consists in the fact that each, in modifying his conduct in accordance with what he thinks the others are going to do, may or may not take account of the effect upon their prices and policies of his own prospective acts. There is an "equilibrium point," such that neither of two sellers can, by changing his price, increase his rate of profit while the other's price remains unchanged. However, if one seller increases his price moderately, thus making some immediate sacrifice, the other will find his most profitable course to lie in increasing his own price; and then, if the original increase is not too great, both will obtain higher profits than at "equilibrium." But that the tendency to cut prices below the equilibrium is less important than has been supposed is shown in the article just referred to.

With an exhaustible supply, and therefore with less to lose by a temporary reduction in sales, a seller will be particularly inclined to experiment by raising his price above the theoretical level in the hope that his competitors will also increase their prices. For the loss of business incurred while waiting for them to do so he can in this case take comfort, not merely in the prospect of approximating his old sales at the higher price in the near future but also in the fact that he is conserving his supplies for a time when general exhaustion will be nearer and even the theoretical price will be higher. Thus a general condition may be expected of higher prices and lower rates of production than are given by the solution of the simultaneous characteristic equations.

For complementary products, such as iron and coal, the situation is in some ways reversed. Edgeworth in his *Papers Relating to Political Economy* points out that when two complementary goods are separately monopolized the consumer is worse off than if both were under the control of the same monopolist. This assumes the equilibrium solution to hold. The tentative deviations from equilibrium made in order to influence the other party may now be in either direction, according as the nature of the demand function and other conditions make it more profitable to move toward the lower prices and larger sales characterizing the maximum joint profit, or to raise one's price in an effort to force one's rival to lower his in order to maintain sales. When the supplies of the complementary goods are exhaustible, the same indeterminateness exists.

A very different problem of duopoly involving the calculus of variations has been studied by C. F. Roos,[14] who finds[15] that the respective profits take true maximum values. However, as in the static case, no definitively stable equilibrium is insured by the fact that each profit is a maximum when the other is considered fixed, since the acts of one competitor affect those of the other. The calculus of variations is used by Roos and G. C. Evans[16] to deal with cost and demand functions involving the rate of change of price as well as the price. Such functions we have for concreteness and simplicity avoided, but if they should prove to be of importance in mine economics the foregoing treatment can readily be extended to them. (Cf. Section 12.) Evans and Roos are not concerned with exhaustible assets, and assume that at any time all competitors sell at the same price.

The problems of exhaustible resources involve the time in another way besides bringing on exhaustion and higher prices, namely, as bringing increased information, both as to the physical extent and condition of the resource and as to the economic phenomena attending its extraction and sale. In the most elementary discussions of exchange, as in bartering nuts for apples, as well as in discussions of duopoly, a time element is always introduced to show a gradual approach to equilibrium or a breaking away from it. Such time effects are equally or in even greater measure involved in exploiting irreplaceable assets, entangling with the secular tendencies peculiar to this class of goods. With duopoly in the sale of exhaustible resources the possibilities of bargaining, bluff, and bluster become remarkably intricate.

The periodic price wars which break the monotony of gasoline prices on the American Pacific Coast are an interesting phenomenon. Along most of the fifteen-hundred-mile strip west of the summits of the Sierras a few large companies dominate the oil business. In the southern California oil fields, however, numerous small concerns sell gasoline at cut prices. Cheap gasoline is for the most part not distilled from oil but is filtered from natural gas, and may be of slightly inferior quality; nevertheless, it is an acceptable motor fuel. The extreme mobility of purchasers of gasoline reduces to a minimum

---

[14] "A Mathematical Theory of Competition," *American Journal of Mathematics*, XLVII (1925), 163.

[15] "Generalized Lagrange Problems in the Calculus of Variations," *Transactions of the American Mathematical Society*, XXX (1927), 360.

[16] *A Mathematical Introduction to Economics* (McGraw-Hill, 1930).

the element of gradualness in the shift of demand from seller to seller with change of price. Ordinarily, the price outside of southern California is held steady by agreement among the five or six major companies, being fixed in each of several large areas according to distance from the oil fields. But every year or two a price war occurs, in which prices go down day by day to extremely low levels, sometimes almost to the point of giving away gasoline, and certainly below the cost of distribution. From a normal price of 20 to 23 cents a gallon the price sometimes drops to 6 or 7 cents, including the tax of 3 cents. Peace is made and the old high price restored after a few weeks of universal joy-riding and storage in every available container, even in bath tubs. The interesting thing is the slowness of the spread of these contests, which usually begin in southern California. The companies fight each other violently there, and a few weeks later in northern California, while in some cases maintaining full prices in Oregon and Washington. These affrays give an example of the instability of competition when variations of price with location as well as time complicate commerce in an exhaustible asset.

Stanford University   HAROLD HOTELLING

Edgeworth's Taxation Paradox and the Nature of Demand and Supply Functions[1]

1. A Tax May Result In Buyers Paying Less

That a tax imposed on the seller of a monopolized article may lead to an actual lowering of the price to the buyer has been shown by F. Y. Edgeworth.[2] His example was of a railway supplying two classes of passenger service at different prices and, unhindered by governmental interference, setting its rates so as to make its own profit a maximum. When the company is compelled to pay a tax on each first-class ticket, it finds it profitable, in Edgeworth's example, to reduce rates on both classes of accommodations. Regarding this paradoxical conclusion, Professor Seligman writes:[3]

> The mathematics which can show that the result of a tax is to cheapen the untaxed as well as the taxed commodities will surely be a grateful boon to the perplexed and weary secretaries of the treasury and ministers of finance throughout the world!

Edgeworth regarded the diminution of price as a result of a tax as a phenomenon of monopoly. We shall see that even under competition, whether "free" competition or of the duopolistic variety, a decrease of prices may result from a tax on the sellers.

It must not, of course, be supposed that the lowering of prices with imposition of a tax will occur in all cases; this would not only fly in the face of common sense (which is singularly unreliable in such matters) but would be at variance with the case of a monopolist controlling articles for which there is no joint demand or joint cost, where a familiar demonstration shows that the price of the taxed article will not be diminished but will in general increase. For some pairs of related demand functions a tax on one commodity will cause both to increase in price; for others, one commodity will increase and the other decrease in price. To survey the boundaries of the third class, for which both

---

[1] Presented at a joint meeting of the Econometric Society and Section K of the American Association for the Advancement of Science, at New Orleans, January 1, 1932.

[2] *Paper Relating to Political Economy* (London, 1925), I, 132, 143 ff., and II, 401.

[3] *Shifting and Incidence of Taxation* (4th ed., 1921), p. 214 n.

commodities will decrease in price, when the seller pays a tax, is one of our primary objects.

Some contributions to the theory of demand and supply functions, suggested by the study of the conditions necessary for the existence of Edgeworth's phenomenon but really of much broader application, are embodied in Sections 4, 5, and 6.

The fact that Edgeworth's phenomenon, even for the monopolistic case for which he established its possibility, is little known and less understood may perhaps be attributed to the peculiar forms of his demand functions as much as to the fact that the argument is essentially mathematical. It is natural to think of a railway company setting its rates, which we shall denote by p_1 and p_2, and then letting its customers decide how many rides they will take. If we let q_1 be the number of first-class, and q_2 the number of second-class, journeys, it is natural therefore to think of q_1 and q_2 as functions of p_1 and p_2. Edgeworth, however, did not give these functions explicitly, but gave the prices which the company could obtain as functions of q_1 and q_2. To be sure, his equations can theoretically be solved for q_1 and q_2, but they are of such complex forms that the solution in intelligible terms is virtually impossible. In Section 8 we shall give an example in which q_1 and q_2 are simple functions of p_1 and p_2.

Edgeworth used continuous demand functions and supposed the tax infinitesimal. These conventions are very useful, and after the next section we shall adhere to them. But to show that neither is essential, and to illustrate more clearly the phenomenon, we shall first give an example with discontinuous demand functions and a very large tax.

The scope of application is widened, first, by the fact that a change in the cost of production will have the same effect as a tax; and second, by the fact that a bounty many be treated as a negative tax. Thus, a bounty paid to a seller (whether a monopolist or not) in proportion to his sales of an article may lead him to increase the price of this article, and also the prices of other things which he sells. Furthermore, the results derived for a seller's monopoly may be applied, *mutatis mutandis*, to a buyer's monopoly. Finally, the reasoning may all be applied to railway and utility rates, which essentially are taxes. In Section 7 we derive a criterion for obtaining revenue with minimum burden.

2. A Discontinuous Example

A railway company supplies first- and second-class transportation between two points and has the following potential customers:

> Group A, consisting of 450 persons who will ride first class if this does not cost as much as $5.00 above the second-class fare, or second class if the difference is $5.00 or more, but who will not pay more than $12.00 for a first-class or $7.00 for a second-class ticket.

Group B, 40 persons who will ride first class if the fare is $11.00 or less, but whose distaste for the second class is so profound that they will not use it at any price.

Group C, a mass of 900 who will ride second class if this fare is $8.00 or less, but who will not pay appreciably more for a first-class than for a second-class ticket.

Group D, 200 persons who will travel second class if the fare is not more than $6.00, and who will not pay more even for first-class tickets.

Without affecting the principle involved, we assume that there is no cost of production, that the company will set rates which will maximize its profits, and that the prospective passengers will each take one trip, if any, per unit of time. If there is an extra cost to the company, of, say, $10.00, for each passenger, each fare mentioned may be increased by $10.00 and the argument left unaffected.

Under these conditions (in the absence of cost), the first-class rate will be set at $12.00 and the second-class at $8.00. Group A will then ride first class, Group C second class, and Groups B and D not at all. The revenue will be $5,400 + $7,200 = $12,600. It is easy to see that this is the maximum possible revenue, losses being incurred if either fare is lowered or raised.

A tax of $7.00 for each first-class ticket is now imposed upon the railway company, those of the second class not being taxed. Since the maximum net profit from a first-class ticket is now less than the second-class rate which will bring about all possible travel in this class, the company's interests call strongly for the diversion of its first-class riders to second class. This cannot, however, be accomplished without reducing the second-class fare to $7.00 or less, because of the strong preference of Group A for the first class. But if the second-class fare is to go so low, the question comes into view of lowering it still further so as to bring in Group D. At first this reduction of second-class fare from $7.00 to $6.00 in order to gain 200 extra passengers seems unprofitable because of the loss of $1.00 on each of the 1,350 in Groups A and C, while the new business amounts only to $1,200. Thus, a loss of $150 would seem to attend the lowering of this fare from $7.00 to $6.00. However, the 40 persons in Group B, in spite of their dislike of second class, will have an influence upon it. If the first-class fare is lowered to $11.00, out of which the $7.00 tax is paid, Group B will yield a net profit of $160. In order to get this, Class A must be prevented from returning to the first class, and this necessitates the lowering of the second-class fare to $6.00. Thus, the apparent $150 loss from this final lowering of the second-class fare is more than offset by the $160 from Group B.

By thus lowering both its rates, the company obtains a profit of $160 from the first class and $9,300 from the second class, a total of $9,460. At the old $12.00 and $8.00 schedule, Group A would have yielded a net profit of $2,250 and Group C $7,200, a total of $9,450. The imposition of the tax therefore means that the company can, by decreasing both rates, make its loss $10.00 less than it would be at the old rates.

3. IS EDGEWORTH'S PHENOMENON IMPROBABLE?

The construction of demand functions displaying the decrease of both prices with the taxation of one commodity has, for the case of monopoly, been difficult and laborious. Many examples have been tried, both by the present author and by others, with the confident expectation of illustrating Edgeworth's phenomenon, only to prove in the end to be unsuitable. This tends to suggest that the decrease of price with taxation may not be very common. However, the failure of the attempted examples was most frequently in the contravention by the proposed demand functions of some condition which demand functions ought to satisfy, rather than in failure to produce decreases of price with the tax.

Thus, if the quantities which will be taken at prices p_1 and p_2 are given by

$$q_1 = 32 - 7p_1 + 16p_2, \qquad q_2 = 32 + 8p_1 - 24p_2,$$

and if the commodity corresponding to the subscript 1 is taxed at rate t, it is easy to see that, if cost of production is negligible, the prices which will maximize the monopoly profit are

$$p_1 = 24 - \frac{t}{2}, \qquad p_2 = \frac{38}{3} - \frac{7t}{12}.$$

Both these prices diminish as the tax increases.

Against this example, which I showed to Professor Holbrook Working several years ago, he made two decisive objections. In the first place, it is unreasonable to suppose that an increase in p_1 should increase q_2 more than it decreases q_1; if railway fares are in question, the increase in the first-class fare will presumably increase second-class travel only by diverting to the second class a part of the first-class business, the remainder perhaps refusing to ride at all. Objections of this kind will be avoided if the conditions to be discussed in Section 5 are satisfied. The other objection is that the maximizing p's make q_2 negative when $t = 0$.

These and other traps must be avoided in the construction of examples. It is therefore necessary, in order to examine Edgeworth's paradox, to determine just what conditions must be satisfied, and also what conditions are likely to be satisfied, by demand functions. This we shall undertake in Section 5. Indeed, as Edgeworth remarked, the chief use of paradoxes is to stimulate a more critical examination of what is already supposed to be known. It is easy to fall into the error of supposing that demand functions for several commodities need satisfy no conditions except the decrease of demand for each commodity when its price increases.

In addition to the conditions which demand functions must necessarily satisfy, there are others which they may be expected to satisfy approximately.

These, which may help to answer the question arising in statistical inquires as to the shape of the demand curve for a single commodity, will be discussed in Section 4.

Edgeworth was interested in the question whether demand functions leading to the phenomenon he discovered are *probable*. In the complete absence of empirical evidence on this point, the only possible type of investigation seems to consist in determining whether, for demand functions involving a finite number of parameters, the condition that Edgeworth's situation shall obtain involves a decrease in the number of parameters. If there are k parameters in a set of demand functions for any number of commodities, each of these sets of functions may be represented by a point in a k-dimensional region. A portion of this region corresponds to demand functions such that the imposition of a small tax on a particular commodity will result in a diminution of the prices of all the commodities considered. It is natural to ask: How large a fraction of the whole region is occupied by this subregion? Unfortunately, no consistent answer is possible, because of the lack of uniqueness in the choice of the parameters; that is, if we should try to compare the k-dimensional volume of the subregion with that of the whole region, we should have to face the possibility of a transformation of the parameters which should change the ratio. But if the region occupied by the points corresponding to demand functions possessing Edgeworth's property is of fewer than k dimensions, their probability may be said to be zero.

Thus, the only answer which at present can confidently be given to the question of the importance in actual economic life of the kind of situation portrayed by Edgeworth turns upon the question whether the number of degrees of freedom of demand functions is decreased when the requirement of lowering of prices with taxation is imposed. We shall see that this is not the case. Not equations, but inequalities, express the phenomenon; and these, if consistent, do not reduce the number of degrees of freedom. There is no basis known at present for denying that Edgeworth's phenomenon may pertain to a large proportion of ordinary situations, or for affirming that it is, in his language, a mere *curiosum*. It is quite possible, so far as the evidence now goes, that in many practical questions of governmental policy the best expert advice has gone astray because of reliance on the simplified cases treated in the textbooks, in which the correlation of demand for different commodities is neglected. Factual investigations of particular sets of commodities may some day produce more definite knowledge of the effects of taxes upon prices. For a random case, the purely deductive reasoning now available fails to tell us definitely whether it is more probable that a tax will increase or decrease the price paid by buyers.

In order to make sure that we have examined all the limitations which may reasonably be applied to demand and supply functions in general, we shall digress to consider some general properties, chiefly of interest on their own account rather than for their bearing upon Edgeworth's phenomenon.

4. THE NATURE OF SUPPLY AND DEMAND FUNCTIONS FOR SINGLE COMMODITIES

It has often been said that choosing a type of curve to fit to data is a more serious statistical problem than the process of fitting itself. In the statistical determination of demand and supply curves it may therefore be worth while to observe that these curves essentially represent cumulated frequency distributions. They are simply ogives.

To examine the histogram corresponding, for example, to a supply function, we may plot the price p horizontally (Figure 1) and take as ordinate the value of the *slope* $q' = dq/dp$ of the supply curve. The area inclosed by this (the unbroken) curve, the p-axis, and any two ordinates represents the additional amount Δq which will be put on the market when the price is increased from the abscissa of the left-hand vertical line to the abscissa of the other vertical line. This increase of supply Δq is also represented by the difference between the corresponding ordinates of the dotted supply curve. Here we are dealing with a long-run effect, a matter of economic statics, not with transitional phenomena. In speaking of increases of price and quantity, we merely contrast two hypothetical permanent situations.

Now the frequency distributions of most important variates have certain properties arising from the influence upon the variate of many causes which are to some extent independent of one another. Usually there is a concentration about a central value, with a rather smooth decline on each side of this value. Commonly there appears to be contact of high order with the axis at each end. Very often there is a rough approximation in form to the normal error curve. There are good reasons to expect these properties. Indeed, some of the causes acting will tend to make the variate large, and some will tend to make it small, in far more cases than those in which all the causes happen to act in the same direction.

Consider, for example, the variation of the amount of milk brought into

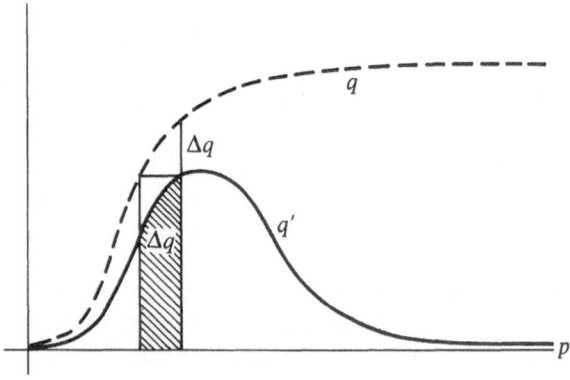

FIGURE 1

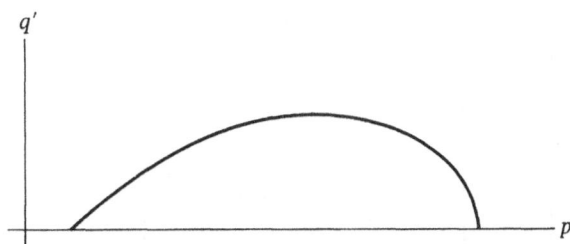

FIGURE 2. Marginal milk production corresponding to change of area. First approximation to differential supply curve.

a city in relation to the price. An increase in the city price will tend to increase production by extending the area from which milk is shipped, by diverting to milk production a larger proportion of the area within the region from which milk is shipped, by encouraging the use of more labor and feed in the better care of cows, and in other ways. Let us consider first the increase of production by extension of area, supposing the simplest possible case, with transportation cost proportional to straightline distance. For very small prices, milk can come only from a small circle about the city. The marginal increase dq in quantity resulting from an increase dp in the price will come from a strip of fixed breadth running around the circle. The area of this narrow strip, and therefore dq, will be proportional to the radius, and therefore to the excess of p above the minimum price at which milk can be produced. If, therefore, we plot $q' = dq/dp$ as a function of p, and suppose that all factors other than producing area remain constant, we shall have a graph (Figure 2) which rises in a linear fashion from the left. It cannot rise indefinitely with increase in price, since the total area from which it is physically possible to ship milk is limited. If larger and larger concentric circles about the city are drawn, a stage will be reached at which only a portion of the circumference lies within a potential milk-producing area. Corresponding to this, our graph will, as we proceed from the left, rise less rapidly. It will at last drop gradually to zero.

Within a particular zone about the city, the land which will be devoted to milk production will depend upon the price in a manner which may perhaps be illustrated by Figure 3, in which the ordinate represents the differential increment in the shipment of milk from this zone on account of the increase in dairy land corresponding to a fixed small increment of price, added to the variable price represented by the abscissa. In order to display the approach to normality with compounding, we have arbitrarily represented this distribution as of a form much more removed from that of the normal curve than would probably be the case.

Now if this or some similar distribution of land in dairying in relation to price be attributed to each zone, and if we consider the variation of milk shipments resulting from the *two* variables—total area, and proportion of land in dairying—we shall obtain a new curve of distribution (Figure 4), which

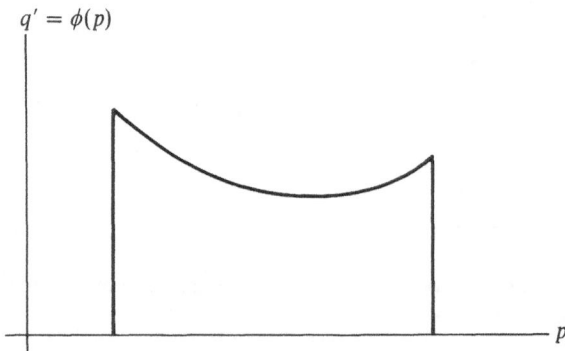

FIGURE 3. Marginal milk production within a particular zone due to change in land devoted to dairying.

may be derived mathematically from those of Figures 2 and 3. We shall now show that the relationship is identical with that between the frequency distribution of the sum of two variates and the separate frequency distributions of these variates.

Let the equation of the curve in Figure 2 be $q' = f_1(p)$. If, in accordance with our first approximation, which supposes 100 per cent of the land in the region devoted to dairying, we suppose unit quantity of milk produced on each unit of land, and if, also, the transportation cost per unit distance is unity, this means that $f_1(r)$ is the length of the arc within the possible milk-producing region of a circle of radius r.

As a second approximation, let us suppose that within each zone a fraction $\phi(p_1)$ of the land will be devoted to dairying when the net price obtainable, after paying for transportation, is p_1. Then, since $p_1 = p - r$, the total amount of milk produced will be

$$f_2(p) = \int f_1(r)\phi(p - r) \, dr,$$

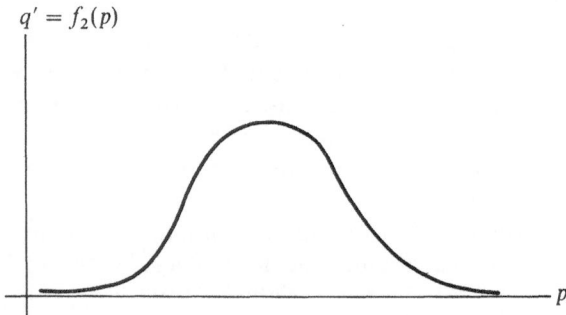

FIGURE 4. Second approximation to the differential supply curve.

the integration being over the range for which both factors are positive. But this is exactly the formula for finding the frequency function $f_2(p)$ for the sum p of two independent variates, r and p_1, whose respective frequency functions are $f_1(r)$ and $\phi(p_1)$.

The resulting frequency function $f_2(p)$ (Figure 4) will be the derivative of the second approximation to the supply curve. It will have contact with the p-axis of higher order than either of the curves of Figures 2 and 3 from which it was obtained. The general theory of the compounding of frequency functions indicates that in other ways also it will resemble the normal frequency curve more than do the distributions which give rise to it.

If we consider a third variable cause of milk production which responds to price, such as the amount of labor put into the care of the cows, we may obtain similarly a third approximation $f_3(p)$ to the differential supply curve; and this will still more closely resemble the normal error curve. There is, to be sure, an assumption analogous to that of independence in the probability theory. This assumption, in obtaining the second approximation, is that the distribution $\phi(p_1)$ is the same for all zones; for the third approximation, it would be assumed that any given residuum of price left after paying for transportation and rent would have an effect in inducing the expenditure of additional labor which is felt in equal proportions in each of the marginal productions previously dealt with. However, the limitation upon the generality of the conclusion that the normal error curve is approached is not very seriously limited by this consideration. Indeed, the work of Markoff, S. Bernstein, and others has shown clearly that the limit theorem of probability remains true in spite of very considerable relaxations of the requirement of independence.

Perhaps a more serious difficulty is in the requirement of the limit theorem of probability that the constituent variables shall have variances of the same order of magnitude. How much this will modify the conclusion that something like the normal distribution is approached must be left for investigation in individual cases. Certainly it will not detract from our confidence in the fundamental proposition that the curve of differential supply as a function of price will have contact of high order with the axis of price at the ends of its range. The exact location of the limits of price outside of which no change of price can affect the supply is unimportant. In many cases no great error will be made if the range is taken as infinite.

The supply curve usually discussed will thus have a shape which can be determined in its general outlines on the basis of pure theory. If p is measured horizontally, this curve may be expected to rise from zero more and more rapidly until a point of inflection is reached; thereafter it will approach asymptotically, or else attain, a horizontal level which it cannot exceed. The ideal curve will satisfy the differential equation

$$\frac{dq}{dp} = e^{-(ap^2 + 2bp + c)}.$$

The argument holds similarly for demand functions. As a price diminishes, different types of demand are disclosed. Cheap wheat is fed to chickens, cheaper wheat to cattle, and still cheaper wheat to stoves. For any one time, however, the price of wheat varies greatly in different places. If at different places there is the same demand function for local consumption in terms of the local price, the total demand will be represented by a compound obtained by methods similar to those used above for supply functions. A curve of differential demand may therefore be expected to have something of the form of the normal error curve, while the demand curve in the ordinary sense, representing the area under the differential demand curve to the left of a particular ordinate, will decline in such a way that the approach to zero for very high prices is extremely gradual. We cannot, of course, assert that the integral of the normal probability curve will fit every supply or demand curve closely; but the reasons just reviewed seem to raise a very considerable presumption of an approximate fit.[4]

The integrated normal distribution may thus reasonably be used to represent both supply and demand functions. This function may indeed find many applications in economics. The trend of a time series may perhaps be somewhat of this character. A historic event (for example, an invention) causes an increase in an economic variable; but this does not take place all at once. The old methods survive in different places and in varying degrees for some time. Gradually they give way, and the curve of sales of the new commodity goes up. The change takes place at first slowly, then more rapidly, and finally, when it is almost complete, slowly again. The observed fact that the logistic, or hyperbolic tangent curve, which is of much the same form as the integrated normal curve, approximates many series suggests that something of this sort represents the rationale of economic and social changes.

5. Demand and Supply Functions for Several Commodities

The quantities which buyers will take of n commodities when the prices of these commodities are $p_1, \ldots, p_n$ are thought of as given by n expressions

$$q_i = F_i(p_1, \ldots, p_n) \qquad (i = 1, 2, \ldots, n). \tag{1}$$

In general it may be assumed that these equations are capable of solution, giving for the prices

$$p_i = f_i(q_1, \ldots, q_n) \qquad (i = 1, 2, \ldots, n). \tag{2}$$

---

[4] Since the foregoing theoretical derivation of the form of a supply curve was written, I learn that curves resembling this general form have been obtained statistically for potatoes and for cabbages by Dr. Louis H. Bean in the *Journal of the American Statistical Association*, XXV (1930), 428–39, and in the *Proceedings of the Seventeenth Annual Meeting of the Potato Association of America* (December, 1930), pp. 53–61. Professor Henry Schultz, to whom I am indebted for remarking on the work of Bean, also observes, apropos the next paragraph, that the integral of the normal curve was used as a trend for the first time by the late Professor R. A. Lehfeldt ("The Normal Law of Progress," *Journal of the Royal Statistical Society*, LXXIX [1916], 329–32). Lehfeldt fits the integrated normal curve to British population and trade and to German birth-rates. Another method of deriving equations of demand functions by means of probability is given by T. H. Brown in the *Journal of the American Statistical Association* for June, 1925.

Our present concern is to find what conditions will be satisfied by these functions.

Consider first an entrepreneur. By using quantities $q_1, \ldots, q_n$ of the several commodities (in which term we include, for convenience, services), he succeeds in making sales amounting, in money, to

$$u = u(q_1, \ldots, q_n).$$

Since he spends $p_1 q_1 + p_2 q_2 + \cdots + p_n q_n$, his net profit will be

$$u - p_1 q_1 - p_2 q_2 - \cdots - p_n q_n.$$

If he regards the prices of the articles he buys as fixed, he will therefore make his profit a maximum if he chooses the q's in such a way that

$$\frac{\partial u}{\partial q_i} = p_i \qquad (n \text{ equations}; \quad i = 1, 2, \ldots, n) \tag{3}$$

provided the inequalities on the second derivatives which assure a true maximum are satisfied.

Not only entrepreneurs, but also purchasers of goods for their own consumption, tend to proceed according to the same principle. To be sure, the buyer may not have in mind a quantity so definite as net profit to maximize, and may behave in an irregular and inconsistent manner. But so far as he weighs in a consistent manner the relative utilities of various goods, considering whether one prospective purchase is "a better bargain" than another, it is clear that there is a function u of the quantities purchased which he is trying to make a maximum. For each such function, (3) and the appropriate inequalities will hold. Consequently, (3) will hold when the total of the u's and the total consumption of q_i's are substituted. The adjustment of prices according to the market means that the p's in (3) may be regarded as functions of the q's, in spite of individual buyers considering the p's fixed. Hence, upon differentiating (3) with respect to q_j, interchanging i and j, and subtracting, we obtain

$$\frac{\partial p_i}{\partial q_j} = \frac{\partial p_j}{\partial q_i}, \tag{4}$$

which must be satisfied by (2) for every combination of i and j, if we neglect irrational and inconsistent purchasing and cases in which the buyer cannot find a definite criterion for distinguishing comparative values. It may be remarked that the equations (4) are a sufficient, as well as a necessary, condition for the existence of a function u satisfying (3), and are therefore known as "integrability conditions" for the functions (2).

A true maximum, rather than a minimum or a minimax, will occur if the quadratic form

$$\sum \sum \frac{\partial^2 u}{\partial q_i \, \partial q_j} x_i x_j \tag{5}$$

is negative definite; that is, if it takes only negative values whenever real numbers, not all zero, are put for the x's. This will be true if

$$\frac{\partial p_i}{\partial q_i} < 0, \qquad \frac{\partial(p_i, p_j)}{\partial(q_i, q_j)} > 0, \qquad \frac{\partial(p_i, p_j, p_k)}{\partial(q_i, q_j, q_k)} < 0, \ldots, \qquad (6)$$

the second and later of these quantities being Jacobian determinants, whose signs are to alternate.

The first of the inequalities (6) expresses the familiar property of demand curves of sloping downward. The others are not so familiar, but are similar to a set of conditions on a utility function obtained by V. Pareto in the *Encyclopédie des sciences mathématiques*, Tome I, Vol. IV, Fasc. 4.

If we had set out to make a utility function u of the q's a maximum subject to the condition that the total expenditure

$$p_1 q_1 + p_2 q_2 + \cdots + p_n q_n$$

must have a constant value, we should have obtained, instead of (3),

$$\frac{\partial u}{\partial q_i} = \lambda p_i, \qquad (3')$$

where λ, the "marginal utility of money," is a function of the q's. This proportionality of prices to marginal utilities is familiar. The argument given above attains a more specific result, obviating the factor λ by restricting attention to those cases in which money is spent, as the saying is, to make money. This category of expenditures is very large, including not only all the articles commonly classed as producers' goods but such items as the haircuts and trousers-pressing paid for by the salesman in order to increase his sales. It includes even a good deal of food, clothing, and entertainment which are consumed, at least in part, with a view to making more money.

It is only when the money expenditure is absolutely fixed that the factor of proportionality λ comes into play. In such cases the relations (3) and (4) may not hold accurately. In place of (4) there will be the looser conditions

$$p_k \left(\frac{\partial p_i}{\partial q_j} - \frac{\partial p_j}{\partial q_i} \right) + p_i \left(\frac{\partial p_j}{\partial q_k} - \frac{\partial p_k}{\partial q_j} \right) + p_j \left(\frac{\partial p_k}{\partial q_i} - \frac{\partial p_i}{\partial q_k} \right) = 0. \qquad (4')$$

These are satisfied if (4) is satisfied, but the converse is not necessarily true. The second-order conditions will also be modified.

For the purposes of the present paper we prefer to consider functions satisfying the stricter conditions (3). If cases of Edgeworth's phenomenon can be found for them, such cases belong *a fortiori* to the wider class also.

The difference between the conditions represented by (3) and by (3') is brought out by the case of a rising demand curve for bread sometimes supposed to occur. A rise in the price may so impoverish the poorer classes that they cannot afford other more costly foods and must eat more bread. The inverted form of the demand curve is explained by variation of the marginal utility of money λ. We are not here inquiring whether such conditions exist;

we are only insisting that, as they are atypical, we shall not depend upon them in giving examples of Edgeworth's taxation phenomenon. But for the great bulk of commercial transactions it may be assumed that the buyer will increase the total of his purchases if, by so doing, he can increase his profits. In such cases (3) and (4), not merely (3') and (4'), will be satisfied.

Our preference for (3) and (4) over (3') and (4') may be explained by imagining a milling company whose stockholders have delivered this mandate to the management: "You shall spend exactly one million dollars during the coming year. You may apportion this as you please in purchasing the various cereals and in other expenses. We hope you will make us as large a profit as possible, but we absolutely insist that the total of the checks you write shall not be more nor less than one million dollars." The management of this company is in the same position as an "economic man" trying to make his ophelimity u, a function of $q_1, \ldots, q_n$, a maximum, while keeping $p_1 q_1 + \cdots + p_n q_n$ fixed. That is the problem which gives rise to (3') and (4'). Apart from rentiers, most buyers, even in their personal transactions, resemble our hypothetical stockholders less than the more common stockholder, who is far better pleased if the net profit $u - \sum pq$ (where u represents gross receipts and $\sum pq$ expenses) is a maximum than if $\sum pq$ is restricted by some other consideration.

The derivatives of the q's with respect to the p's obtained from (1) will also, in the wide class in which λ is constant, satisfy relations analogous to (4) and (6). Indeed, it is known[5] that $\partial q_i / \partial p_j$ is the ratio of the cofactor of $\partial p_j / \partial q_i$, in the determinant of the derivatives such as the latter, to the value of this determinant. Since, by (4), this determinant is symmetric,[6] it follows that

$$\frac{\partial q_i}{\partial p_j} = \frac{\partial q_j}{\partial p_i}. \tag{7}$$

Just as we have a utility (or profit) function u of the quantities consumed whose derivatives are the prices, there is, dually, a function of the prices whose derivatives are the quantities consumed. The existence of such a function,

[5] Cf., for example, Eisenhart's *Riemannian Geometry*, p. 2, or books on advanced calculus.

[6] As this paper goes to press, I learn from Professor Henry Schultz, to whom I am grateful for several other comments and suggestions also, that statistical demand equations for agricultural commodities have been worked out in his laboratory, which do not satisfy the integrability conditions (7). Indeed, he states, $\partial q_i / \partial p_j$ differs in several instances from $\partial q_j / \partial p_i$, not only in absolute value but in sign, and with sufficiently small standard errors. His equations involve time explicitly, e.g., $q_i = F_i(p_1, \ldots, p_n, t)$, but the additional variable t should not in itself spoil the integrability conditions if the situation is regarded as a succession of static states. If these results are definite enough to overcome the treacherous difficulties of inference from time series, they constitute a most striking phenomenon, for which some explanation is required. The difference in sign is particularly remarkable. It is as if lower bus rates should increases travel in the same direction on a railroad, though lowering the railroad rate takes away business from the buses. Perhaps consumers in buying these commodities behave in a very different way from what they would if their knowledge and reasoning power were perfect; or changes in fashion, or some other dynamical effect, may have obscured the situation.

which heretofore does not seem to have been noticed, is assured by (7). On the basis of physical analogies we may call this the "price potential."

From the values of the derivatives of the p's with respect to the q's alluded to above, it is possible, with the help of the theorem on minors of adjoint determinants,[7] to express the Jacobians of the p's with respect to the q's in terms of those of q's with respect to the p's. It will then follow from (6) that

$$\frac{\partial q_i}{\partial p_i} < 0, \qquad \frac{\partial(q_i, q_j)}{\partial(p_i, p_j)} > 0, \qquad \frac{\partial(q_i, q_j, q_k)}{\partial(p_i, p_j, p_k)} < 0, \ldots. \tag{8}$$

A consequence of these relations is that if the price potential be expanded about any set of values of the prices, the terms of second degree form a negative definite quadratic form. Thus, for two commodities, if the price potential be thought of as the height of a surface above the plane of p_1 and p_2, the surface will be concave downward. A utility surface similarly plotted over the plane of q_1 and q_2 will likewise be concave downward. The meaning of this criterion has not always been realized. It is not satisfied by one of the utility surfaces depicted by Professor Irving Fisher,[8] a surface which can only represent a state of affairs in which a person may be so oversupplied with a commodity that he will pay to have it taken away.

Let us revert to the hypothetical demand functions mentioned in Section 3, which are of the form

$$q_1 = a - bp_1 + cp_2, \qquad q_2 = a' + b'p_1 - c'p_2,$$

with a, b, c, a', b', c' positive and $b < b'$. Professor Working's objection that an increase in p_1 should not increase q_2 more than it decreases q_1 is applicable to the special case of passenger tickets but not to commodities in general, for in general we could change the units of measure so as to make $b > b'$. Expressing the quantity and price of the first commodity in terms of a new unit by Q_1 and P_1, let us suppose the ratio of units such that $q_1 = hQ_1$. Then, since $p_1 q_1$ must equal $P_1 Q_1$, we shall have $p_1 = P_1/h$. The transformed demand functions are:

$$Q_1 = \frac{a}{h} - \left(\frac{b}{h^2}\right) P_1 + \left(\frac{c}{h}\right) p_2,$$

$$q_2 = a' + \left(\frac{b'}{h}\right) P_1 - c' p_2.$$

If the new unit is such that the criterion suggested by Working is satisfied, we must have

$$\frac{b}{h^2} > \frac{b'}{h}.$$

But the same criterion can be applied to show that, in the same units, an

---

[7] M. Bocher, *Introduction to Higher Algebra*, p. 31.

[8] *Mathematical Investigations in the Theory of Value and Prices*, p. 70, Fig. 19.

increase in p_2 should affect q_2 more than Q_1; whence

$$c' > \frac{c}{h}.$$

Since all the quantities appearing in these two inequalities are positive, we may multiply them member for member, obtaining

$$bc' - b'c > 0.$$

But this is precisely the second of the conditions (8), which in view of (7), may be written

$$\frac{\partial q_i}{\partial p_i} \frac{\partial q_j}{\partial p_j} - \left(\frac{\partial q_i}{\partial p_j}\right)^2 > 0. \tag{9}$$

and which may thus be derived without reference to a utility function, since an arbitrary pair of demand functions may for this purpose be treated as linear. Conversely, if (9) is satisfied, values of h can be found which satisfy the preceding inequalities, and so escape Working's objection.

The foregoing results have been deduced on the assumption of free competition among buyers. If we assume free competition among sellers, we shall come to analogous conclusions regarding supply functions. At prices $p_1, \ldots, p_n$ the amounts which will be put on the market may be written

$$q_i = G_i(p_1, \ldots, p_n) \qquad (i = 1, 2, \ldots, n). \tag{10}$$

Solving these equations,

$$p_i = g_i(q_1, \ldots, q_n). \tag{11}$$

Let $v(q_1, \ldots, q_n)$ be the cost of production of a seller who produces quantities $q_1, \ldots, q_n$. He will try to maximize his profit, which is

$$p_1 q_1 + p_2 q_2 + \cdots + p_n q_n - v. \tag{12}$$

Because free competition exists, he will regard the prices as fixed. Consequently,

$$\frac{\partial v}{\partial q_i} = p_i,$$

so that the integrability conditions (4) hold also for the supply functions (11). Consequently, the integrability conditions (7) hold for the supply functions (10).

A true maximum of (12) will be assured if the quadratic form analogous to (5), with the second derivatives of v replacing those of u, is *positive* definite. This is equivalent to the inequalities

$$\frac{\partial p_i}{\partial q_i} > 0, \qquad \frac{\partial(p_i, p_j)}{\partial(q_i, q_j)} > 0, \qquad \frac{\partial(p_i, p_j, p_k)}{\partial(q_i, q_j, q_k)} > 0, \ldots, \tag{13}$$

which must be satisfied by the supply functions. The first of these is the familiar proposition that a supply curve slopes upward; the rest are believed to be new.

It is obvious that "free competition," in the sense in which we have used the term, excludes both supply curves which slope downward and demand curves which slope upward in the neighborhood of the solution.

In the same way that (8) follows from (6), it follows from (13) that the quantities analogous to those of (8), but calculated from the supply instead of from the demand functions, will all be positive.

These results hold not only for each seller but for the aggregate of sellers, analogously to the case of demand. It also follows that there is a price potential, a function whose derivatives with respect to the prices are the quantities which will be supplied. For two commodities, this sellers' price potential and the cost function may be represented by surfaces which will be concave *upward*.

The equilibrium condition under free competition will be represented by a solution of the simultaneous equations (1) and (10). For two commodities this may be represented graphically by drawing over the plane of q_1 and q_2 the utility surface, which is concave downward, and the cost surface, which is concave upward. The equilibrium q's are the co-ordinates of a point through which a vertical line cuts the two surfaces in points the tangent planes at which are parallel. A similar graphic interpretation may be made by means of the two price potentials.

6. SUGGESTIONS FOR STATISTICAL STUDIES OF DEMAND AND SUPPLY

The statistical determination of demand and supply functions has hitherto been undertaken for single commodities and has encountered in most cases wide fluctuations which lead to great uncertainty. A more hopeful approach may perhaps involve the simultaneous study of groups of related commodities. In such statistical studies, the integrability conditions (4) and (7) should receive careful consideration. In fitting a linear demand function, for example, to the prices and consumptions of n commodities, there are n^2 coefficients to be determined, in addition to the n means. However, the use of the integrability conditions reduces the number of independent coefficients from n^2 to $n(n + 1)/2$, with a corresponding gain in the accuracy of the determination by the reduction in the number of degrees of freedom absorbed by these constants. Or the n^2 coefficients might be determined as independent quantities statistically, and then the differences of those symmetrically placed might be compared with their standard errors. In this way a test of the validity of the integrability conditions as applied to the commodities in question could be made. The difference of two symmetrically placed coefficients could be taken as a measure of the degree of inconsistency in buyer's judgments, or of the rigidity of an absolute limit on their money expenditures which cannot be increased with the help of the efficiency to be gained by increased purchases.

When such statistical studies have progressed far enough to consider deviations from linearity, it will be interesting to see whether functions occur which generalize the integrated normal distribution of Section 4. For such functions,

$$\log\left(\pm\frac{\partial q_i}{\partial p_j}\right) = -\sum_k \sum_l a_{ijkl} p_k p_l + 2\sum_k b_{ijk} p_k + c_{ij},$$

the ambiguous sign being $+$ for supply and $-$ (minus) for demand functions when $i = j$. When $i \neq j$, the sign will depend upon whether the commodities compete with or complement each other.

Measures of "partial elasticity of demand" generalizing Marshall's definition of elasticity have been proposed by Professor Henry L. Moore,[9] who has applied to them a notation suggested by Yule's notation for partial correlation. The range of possible values of these coefficients is infinite. I venture to suggest another measure of correlation of demand which may sometimes prove useful, namely,

$$c_{ij} = \frac{\dfrac{\partial q_i}{\partial p_j}}{\sqrt{\dfrac{\partial q_i}{\partial p_i}\dfrac{\partial q_j}{\partial p_j}}}. \tag{14}$$

If c_{ij} is positive, the commodities compete with each other; if it is negative, they are complementary.

The integrability condition (7) gives $c_{ij} = c_{ji}$. Also, $c_{ii} = -1$. From (9) it follows that the value of c_{ij} lies between -1 and 1. Also, c_{ij} is independent of the units of measure and of money. The resemblance of c_{ij} to the familiar statistical correlation coefficient thus suggested is another of those subtle analogies of economics with the theory of probability to which Edgeworth was so fond of alluding. The analogy might be pushed farther by considering functions of the c_{ij} similar to partial and multiple correlation coefficients.

For a system of only two commodities, let us denote by J the Jacobian of the q's with respect to the p's. Then,

$$\frac{\partial q_1}{\partial p_1} = J\frac{\partial p_2}{\partial q_2}, \qquad \frac{\partial q_1}{\partial p_2} = -J\frac{\partial p_2}{\partial q_1}, \qquad \frac{\partial q_2}{\partial p_2} = J\frac{\partial p_1}{\partial q_1}. \tag{15}$$

Hence, for this case,

$$c_{12} = \frac{-\dfrac{\partial p_2}{\partial q_1}}{\sqrt{\dfrac{\partial p_1}{\partial q_1}\dfrac{\partial p_2}{\partial q_2}}}, \tag{16}$$

illustrating again the duality of price with quantity. However, the corresponding expression for systems with more than two commodities is more complex.

Either (14) or (16) may equally be applied to measure correlation of supply, or extent of joint cost, the derivatives being computed in this case from the

[9] *Quarterly Journal of Economics*, XL (1926), 393.

supply or cost functions instead of the demand functions. If the commodities i and j are produced jointly, like ham and bacon, the correlation c_{ij} of cost will be positive. It will be negative if they compete for productive facilities, as wheat and rye compete for the use of fields and farm labor. But the cost correlation c_{ii} is $+1$.

7. TAXATION UNDER FREE COMPETITION

Let $h_i(q_1, \ldots, q_n)$ be the excess of demand price over marginal cost for a fixed set of q's. Then in the notation of (2) and (11),

$$h_i = f_i - g_i.$$

We shall denote differentiation with respect to q_j by an added subscript j, so that

$$h_{ij} = \frac{\partial h_i}{\partial q_j} = \frac{\partial h_j}{\partial q_i},$$

the last equality being a consequence of the integrability conditions (4), which are applicable to both the supply and demand functions. Let

$$D = \frac{\partial(h_1, \ldots, h_n)}{\partial(q_1, \ldots, q_n)} = \begin{vmatrix} h_{11} & h_{12} & \ldots & h_{1n} \\ \cdots\cdots\cdots\cdots\cdots\cdots\cdots \\ \cdots\cdots\cdots\cdots\cdots\cdots\cdots \\ h_{n1} & h_{n2} & \ldots & h_{nn} \end{vmatrix}. \tag{17}$$

Let p_i and q_i be the equilibrium values, for which supply and demand are equal. Then

$$h_i(q_1, \ldots, q_n) = 0 \qquad (i = 1, 2, \ldots, n).$$

Now let a tax t_i per unit sold of the ith commodity be levied upon the producers, and let $p_i + \delta p_i$ and $q_i + \delta q_i$ be the new prices and quantities. (Here i is supposed to take all the values $1, 2, \ldots, n$). The demand price must now exceed by t_i the supply price; hence,

$$h_i(q_1 + \delta q_1, \ldots, q_n + \delta q_n) = t_i.$$

Expanding, and subtracting the previous equation, we have to a first approximation appropriate to small taxes,

$$\sum_j h_{ij}\, \delta q_j = t_i. \tag{18}$$

Solving these equations, we find

$$\delta q_j = \frac{1}{D} \begin{vmatrix} h_{11} & \ldots & t_1 & \ldots & h_{1n} \\ h_{21} & \ldots & t_2 & \ldots & h_{2n} \\ \cdots\cdots\cdots\cdots\cdots\cdots\cdots \\ h_{n1} & \ldots & t_n & \ldots & h_{nn} \end{vmatrix}, \tag{19}$$

the determinant being formed from D by replacing the jth column by the tax-rates.

The price changes to buyers resulting from the taxes are:

$$\delta p_i = \sum_j f_{ij}\,\delta q_j = -\frac{1}{D}\begin{vmatrix} 0 & f_{i1} & \cdots & f_{in} \\ t_1 & h_{11} & \cdots & h_{1n} \\ t_2 & h_{21} & \cdots & h_{2n} \\ \cdots\cdots\cdots\cdots\cdots\cdots \\ t_n & h_{n1} & \cdots & h_{nn} \end{vmatrix}, \tag{20}$$

where $f_{ij} = \partial f_i / \partial q_j$.

In Section 5 it was shown that the derivatives f_{ij} are the coefficients in a negative definite quadratic form and that the g_{ij} are the coefficients in a positive definite quadratic form. Hence the h_{ij} are the coefficients in the difference of these two forms, which is negative definite. It follows that D has the sign of $(-1)^n$.

For a single commodity, $D(= h_{11})$ and f_{11} are negative; hence, (20) gives a proof that the change of price following the imposition of the tax is positive, the familiar result for this case. However, when more than one commodity is involved, this does not hold.

Let us consider two commodities which compete both in production and in consumption, and call the first wheat and the second rye. Suppose that the demand functions are:

$$\left.\begin{array}{l} p_1 = f_1(q_1, q_2) = 41 - 5q_1 - 7q_2 \\ p_2 = f_2(q_1, q_2) = 58 - 7q_1 - 10q_2. \end{array}\right\} \tag{21}$$

The plausibility of these functions will perhaps be more evident if we solve for the q's, obtaining:

$$q_1 = 4 - 10p_1 + 7p_2$$
$$q_2 = 3 + 7p_1 - 5p_2.$$

If in the foregoing equations both quantities be expressed in millions of bushels, and if there is some measurable attribute, such as number of calories, of which a million bushels of wheat possesses 1 and a million bushels of rye possesses 1.4, then the numbers of these caloric units bought in the two forms at given prices per bushel are:

$$q_1' = 4 - 10p_1 + 7p_2$$
$$q_2' = 4.2 + 9.8p_1 - 7p_2.$$

Now, if the price per bushel of wheat increases by unity, that of rye remaining constant, consumers will decrease their consumption of wheat by ten million bushels, that is, by 10 caloric units. Of these, 9.8 caloric units will be made up by increased purchases of rye, while the remaining 0.2 will be dispensed with or obtained from other commodities. Likewise, a unit increase in the price per bushel of rye will mean that 7 fewer caloric units will be derived from rye but will be replaced in the increased consumption of wheat. (These numerical values are, of course, hypothetical, and not of statistical or chemical origin.)

Using millions of bushels, let us assume as supply functions:

$$\left.\begin{aligned} p_1 &= g_1(q_1, q_2) = 13 + 2q_1 + q_2 \\ p_2 &= g_2(q_1, q_2) = 20 + q_1 + q_2. \end{aligned}\right\} \tag{22}$$

Now let taxes t_1 and t_2 per unit quantity produced be imposed upon producers of wheat and of rye, respectively. In view of recent proposals, these taxes might be called "equalization fees." From (20) we find

$$\delta p_1 = \frac{-t_1 + 9t_2}{13}, \qquad \delta p_2 = \frac{-3t_1 + 14t_2}{13}.$$

Putting $t_2 = 0$, these results mean that, if rye is untaxed, a tax of 13 cents a bushel on wheat, if not sufficient to stop wheat production entirely, will enable buyers to get wheat at 1 cent a bushel less, and rye at 3 cents a bushel less, than before. On the other hand, a tax on rye benefits producers of both commodities by raising prices, the price of rye being increased by more than the tax.

Thus the conclusion which Edgeworth reached for monopoly is extended to free competition: a tax on sellers may result in lower prices to buyers, not only of the articles taxed, but of others as well.

The simplicity of the foregoing example makes it possible to examine the effect of increasing the tax. The fact that the buyers' prices (21) must exceed the seller's prices (22) by t_1 and t_2, respectively, yields a pair of linear equations whose solution is

$$q_1 = \frac{4 - 11t_1 + 8t_2}{13}, \qquad q_2 = \frac{42 + 8t_1 - 7t_2}{13}. \tag{23}$$

With (22) these values give as the net prices to the sellers remaining after payment of the tax,

$$p_1 = \frac{219 - 14t_1 + 9t_2}{13}, \qquad p_2 = \frac{306 - 3t_1 + t_2}{13}. \tag{24}$$

Since all these p's and q's must be positive, we have four linear inequalities on t_1 and t_2. If $t_2 = 0$, these give one lower and three upper bounds for t_1. Of the latter, the only one which is effective is the least, namely, 4/11; this may be compared with the price of wheat in the absence of a tax, 219/13. The significance of the lower bound, $-21/4$, is that a bounty on wheat will increase the production of wheat in accordance with the formulas so long as it is less than 21/4; beyond this value the equations will not hold because there is no more rye land which can be transferred to wheat production. But even before these limits are reached, the considerations of Section 4 suggest that the assumed linearity of the functions will cease to be tenable.

We shall now deduce the conditions which must be satisfied for a pair of commodities in order that a tax at rate t on the first may result in lowering both prices under free competition. Since $n = 2$, D is positive. Hence, upon putting $t_1 = t, t_2 = 0$ in (20), the conditions $\delta p_i < 0$ ($i = 1, 2$) become, for $t > 0$,

$$f_{11}g_{22} - f_{12}g_{12} > f_{11}f_{22} - f_{12}^2 \tag{25}$$

$$f_{12}g_{22} - f_{22}g_{12} > 0. \tag{26}$$

Consider the possibility that f_{12} is positive; that is, that the two articles complement each other in consumption. Multiply (26) by f_{12} and (25) by $-f_{22}$, which is necessarily positive, and add. The result, which may be written

$$(f_{12}^2 - f_{11}f_{22})(g_{22} - f_{22}) > 0,$$

is absurd, since the first factor is, by (9), negative, and the second is essentially positive. In case $f_{12} = 0$, the left member of (25) is negative, while the right member is always positive. We conclude that the lowering of the two prices by reason of the tax requires that the commodities compete in consumption. It also requires that they compete in production; for, since we have just seen that f_{12} must be negative, (26) shows that g_{12} is positive.

Let arbitrary negative values be given the f_{ij}'s, subject only to the universal condition that the right member of (25), shall be positive. Also, let an arbitrary positive value be given g_{22}. The inequalities (25) and (26) then merely impose lower bounds on g_{12}. Taking g_{12} great enough to surpass these limits, we may choose for g_{11} a positive value sufficiently great to satisfy the one remaining condition on these quantities, namely, the second of the inequalities (13), which in the notation of the present section becomes

$$g_{11}g_{22} - g_{12}^2 > 0.$$

In this way the first derivatives of the demand and supply functions are selected, without reduction in the number of degrees of freedom in such a way that the tax causes reductions in the prices. The functions themselves may be chosen without further restrictions, except that they must be positive. This verifies for free competition that the phenomenon is not merely possible but, so far as purely logical (not objective) probability goes, is no less probable than the more familiar contrary case.

For a single commodity, uncorrelated with any other, it is well known that an excise tax reduces the total of producers' and consumers' surpluses by more than the revenue received by the state. The total loss of benefit from the commodity which results from the tax is, apart from higher powers, proportional to the square of the tax. We shall now consider how this situation generalizes to a group of commodities.

By the total benefit we mean

$$w = u - v,$$

where u and v have the same meanings as in Section 5 (where they were taken as sums for all persons in the market), from which it follows that

$$\frac{\partial w}{\partial q_i} = h_i. \tag{27}$$

Denoting the values which quantities take in the absence of taxation by enclosing them in parentheses, we have for the total benefit when taxes $t_1, \ldots, t_n$ are levied on the several commodities,

$$w = (w) + \sum \left(\frac{\partial w}{\partial t_i}\right) t_i + \frac{1}{2} \sum \sum \left(\frac{\partial^2 w}{\partial t_i \, \partial t_j}\right) t_i t_j + \cdots. \tag{28}$$

Now

$$\frac{\partial w}{\partial t_i} = \sum_k \frac{\partial w}{\partial q_k} \frac{\partial q_k}{\partial t_i} = \sum_k h_k \frac{\partial q_k}{\partial t_i},$$

by (27). But since, as in the equation following (17), $h_k = 0$ at equilibrium, all the terms of first degree in the tax-rates vanish, as in the case of a single commodity. Differentiating again,

$$\frac{\partial^2 w}{\partial t_i \, \partial t_j} = \sum_k \sum_l h_{kl} \frac{\partial q_k}{\partial t_i} \frac{\partial q_l}{\partial t_j} + \sum_k h_k \frac{\partial^2 q_k}{\partial t_i \, \partial t_j}, \tag{29}$$

the last term dropping out at equilibrium. If we put

$$\left(\frac{\partial q_k}{\partial t_i}\right) = H_{ki},$$

it is evident from (19) that H_{ki} is the cofactor of h_{ki} in D, divided by D. Consequently,

$$\sum_k h_{kl} H_{ki} = \delta_{li},$$

where δ_{li} is the Kronecker delta, equal to unity if $l = i$, and otherwise equal to zero. Hence from (29),

$$\left(\frac{\partial^2 w}{\partial t_i \, \partial t_j}\right) = \sum_l \delta_{li} H_{lj} = H_{ij}.$$

Thus from (28) we obtain, neglecting powers and products of the t's of degrees higher than the second,

$$\delta w = w - (w) = \frac{1}{2} \sum \sum H_{ij} t_i t_j. \tag{30}$$

This is the change in total benefit; it may also be written

$$\delta w = -\frac{1}{2D} \begin{vmatrix} 0 & t_1 & t_2 & \cdots & t_n \\ t_1 & h_{11} & h_{12} & \cdots & h_{1n} \\ t_2 & h_{21} & h_{22} & \cdots & h_{2n} \\ \cdots & \cdots & \cdots & \cdots & \cdots \\ t_n & h_{n1} & h_{n2} & \cdots & h_{nn} \end{vmatrix}.$$

The theorems on adjoint determinants show that, since the h_{ij} are the coefficients of a negative definite quadratic form, the same is true of the H_{ij}. Hence δw is negative; for every set of tax rates there is a loss of total benefit, as measured by the sum of producers' and consumers' surpluses and the state revenue. It follows *a fortiori* that the sum of producers' and consumers' surpluses will be reduced by the taxes—a conclusion not entirely obvious otherwise.

The conclusion that the total benefit will necessarily be diminished by the taxes requires the qualification that we have taken no account of what the government does with the money. If the taxes collected from the kind of commodities we have been considering, produced with increasing cost, are used to subsidize industries operating with decreasing cost, with proper regulation to insure that the subsidy will result in sufficiently lower prices to consumers, there may well be a net gain in benefit. For example, it might be an economic thing to levy an excise tax upon agricultural products in order to bring about lower railway rates by a subsidy. However, we can show that it will not pay to tax some increasing-cost industries to subsidize others. For the negative definite quadratic form in (30) is *ipso facto* negative, whether the t's are positive or negative or some are of each sign; the subsidies are represented simply by negative t's.

If a certain revenue $\sum q_i t_i$ is to be raised by excise taxes, it is desirable to adjust the rates so that the net loss (30) will be a minimum. The solution of this problem is given by

$$\sum_j H_{ij} t_j = \lambda q_i,$$

λ being a constant. Solving these equations,

$$t_j = \lambda \sum h_{ij} q_i.$$

The value of λ is determined by the revenue desired, which is the value obtained by multiplying the last equations by q_j and summing for all the values of j.

If one and only one of the commodities (for example, land) has a supply or a demand independent of its price, the corresponding h_{ii} will be infinite. This occurs in only one of the equations of the last set. Consequently the whole revenue should, if possible, be derived by taxing this one commodity.

It has been shown by various writers that a tariff can be justified from a strictly nationalistic point of view if judiciously levied, provided account is taken of the public revenue obtained. The results of this section show that from the standpoint of this same narrow nationalism the justification can in some possible cases be found without supposing that the revenue exceeds the cost of collecting it, for we have seen that the (foreign) sellers may lower their prices to buyers, in addition to paying the tax. In this proof it has not been necessary to make the usual appeal for diminishing-cost industries or to assume a foreign monopoly.

On the other hand, a tariff, like our hypothetical tax on rye, may result in foreign sellers actually receiving more for their products. In either case, a net damage results from the imposition of the duty when the whole human race is considered.

Free competition, with a single price for each commodity, has been assumed up to this point. It is possible to have as many prices for a commodity as there are sellers, and the number of sellers may be small.[10] It is easy to see that in this case also it is possible that a tax on one of two related articles may cause both prices to be lowered, though the sellers pay the tax. Indeed, cases of this limited competition, or duopoly, can be imagined in which the sellers are so separated geographically, or so different in some other respect, that there is arbitrarily little of the competitive element. In other words, cases of competition may exist which are as much as one pleases like monopoly; and for monopoly the phenomenon can occur, as Edgeworth showed, and as we shall now see in more detail.

8. CONDITIONS FOR EDGEWORTH'S PHENOMENON WITH MONOPOLY

The cost function plays a much less essential part in producing Edgeworth's phenomenon under conditions of monopoly than in free competition. It is no longer necessary to assume that the commodities compete in production in order to find demand functions which will lead to a lowering of price by taxation. In what follows, the costs may be positively or negatively correlated or independent, or they may be zero; diminishing costs may obtain; but the results deduced will, in their general nature, be unaffected. As soon as a cost function is known, it will be possible to find demand functions displaying Edgeworth's phenomenon.

The monopoly profit will be a function of the prices and of the quantities sold; but on account of the demand relations (1) and (2), it may be expressed in terms of the prices alone or of the quantities alone. Merely for the sake of parallelism with our work for free competition, let us regard the quantities as the independent variables, as Edgeworth did. We shall indicate by added subscripts differentiation with respect to the q's having these subscripts. Denoting the monopoly profit in the absence of a tax by π, the first-order conditions that it be a maximum are thus $\pi_i = 0$. The second-order conditions are

$$\pi_{ii} < 0, \qquad \pi_{ii}\pi_{jj} - \pi_{ij}^2 > 0, \dots; \tag{31}$$

these are necessary and sufficient that the π_{ij} be the coefficients in a negative definite quadratic form.

If a tax at rate t is imposed upon the first commodity only, to be paid by the seller, the function to be made a maximum is $\pi - tq_1$. The first-order conditions are now

$$\pi_1 = t, \qquad \pi_2 = \pi_3 = \pi_4 = \cdots = 0. \tag{32}$$

[10] Cf. my "Stability in Competition," *Economic Journal*, XLI (1929), 41.

Denoting derivatives with respect to t by a prime, we have from (32),

$$\sum_j \pi_{ij} q'_j = \delta_{i1}, \tag{33}$$

the Kronecker delta on the right having the value unity when $i = 1$, and being otherwise equal to zero. These equations may be solved for the q'_j. Let

$$\sigma_{ij} = \frac{\text{Cofactor of } \pi_{ij} \text{ in the determinant } |\pi_{ij}|}{\text{Determinant } |\pi_{ij}|}.$$

We may express this relation by saying that σ_{ij} is the conjugate of π_{ij}. It may then be proved that π_{ij} is the conjugate of σ_{ij}. Moreover, the σ_{ij} will satisfy conditions analogous to (31), and so will be the coefficients of a negative definite form.

The solution of (33) may now be written

$$q'_j = \sigma_{1j}. \tag{34}$$

The price changes resulting from the tax will be at the rates

$$p'_i = \sum_j p_{ij} q'_j = \sum_j p_{ij} \sigma_{1j}, \tag{35}$$

with an approximation which is adequate if t is sufficiently small. We are interested in the conditions under which all the quantities (35) are negative.

For the case of two commodities, to which we shall devote the remainder of this section, (35) shows that the tax will diminish the prices if

$$p'_1 = p_{11}\sigma_{11} + p_{12}\sigma_{12} < 0, \tag{36}$$

$$p'_2 = p_{12}\sigma_{11} + p_{22}\sigma_{12} < 0. \tag{37}$$

Since each of the factors in the first term of (36) is negative, we cannot have $p_{12} = 0$. Moreover, p_{12} cannot be positive. For if we suppose that it is, we may multiply (37) by it, and (36) by $-p_{22}$, and add; the result of this would be

$$(p_{12}^2 - p_{11}p_{22})\sigma_{11} < 0,$$

which is absurd, because both factors on the left must be negative. Also, since p_{11}, σ_{11} and p_{12} are negative, it follows from (36) that $\sigma_{12} > 0$.

Thus with monopoly, as with free competition, it is necessary that the two commodities compete for buyers if Edgeworth's phenomenon is to take place. This is expressed by the condition $p_{12} < 0$. In neither case is joint cost sufficient to produce the phenomenon in the absence of substitutability in consumption.

Subject to this one condition, and to the conditions of Section 5 on demand functions in general, we can assign arbitrary values to the first derivatives p_{ij} of the demand functions, and also any cost function, and then determine the second derivatives of the demand functions in such a way that Edgeworth's phenomenon will occur. For, since all the coefficients of σ_{11} and σ_{12} in (36) and (37) are then negative, an arbitrary negative value may be given σ_{11}, and

(36) and (37) merely impose two lower limits on σ_{12}. When σ_{12} is chosen greater than both these limits, σ_{22} can be found to satisfy the remaining necessary condition, $\sigma_{11}\sigma_{22} - \sigma_{12}^2 > 0$. The conjugate quantities π_{ij} may then be calculated, and will satisfy the requisite conditions (31).

The next step in choosing the demand functions involves the cost function v, whose derivatives we shall write g_1 and g_2, as before. Since

$$\pi = p_1 q_1 + p_2 q_2 - v,$$

we have

$$\left.\begin{aligned}
\pi_1 &= p_1 + p_{11}q_1 + p_{12}q_2 - g_1 \\
\pi_2 &= p_2 + p_{12}q_1 + p_{22}q_2 - g_2.
\end{aligned}\right\} \tag{38}$$

Differentiating again,

$$\left.\begin{aligned}
\pi_{11} &= 2p_{11} + p_{111}q_1 + p_{112}q_2 - g_{11}, \\
\pi_{12} &= 2p_{12} + p_{112}q_1 + p_{122}q_2 - g_{12}, \\
\pi_{22} &= 2p_{22} + p_{122}q_1 + p_{222}q_2 - g_{22}.
\end{aligned}\right\} \tag{39}$$

Any positive values may now be given q_1 and q_2. Everything appearing in (39) has now been chosen excepting the four second derivatives $p_{111}, p_{112}, p_{122}$, and p_{222}, which incidentally are the third derivatives of the utility. In (39) we have three linear equations in these four quantities. Their consistency is evident, since p_{111} may be given a completely arbitrary value, and the rest are then determined in turn. The values, corresponding to maximum profit, of the first and second derivatives of the p's having now been chosen, nothing remains but to choose the constant terms in the expansion about the maximizing values, terms which need only to be positive, and the terms of orders higher than the second; these are in no way restricted by the condition that the prices be lowered by the tax.

In this way it is easy to construct demand functions giving the effect under consideration, the p's being obtained explicitly as functions of the q's. The simplest functions of the q's would be polynomials of the second degree; Edgeworth's second example was of this character, his first being more complicated.

A closely similar procedure to that above yields demand functions which exhibit the phenomenon in the somewhat more desirable form of functions of the prices which give the quantities bought. An example with zero cost is

$$\left.\begin{aligned}
q_1 &= 1 - 2(p_1 - 1) + (p_2 - 1) - 30(p_1 - 1)^2 + 7(p_1 - 1)(p_2 - 1), \\
q_2 &= 1 + (p_1 - 1) - 2(p_2 - 1) + \tfrac{7}{2}(p_1 - 1)^2.
\end{aligned}\right\} \tag{40}$$

Unit quantities are sold at unit prices to yield a maximum monopoly profit in the absence of a tax, if there is no cost of production. For a sufficiently small tax on the first commodity, both prices decrease. To verify this, we note that the monopoly profit is

$$\pi = p_1 q_1 + p_2 q_2 - tq_1. \tag{41}$$

Denoting temporarily differentiation with respect to p_1 and p_2 by subscripts, we have as the conditions for a maximum,

$$\pi_1 = 0, \qquad \pi_2 = 0.$$

These equations hold for every value of t. We may therefore differentiate them with respect to t. This gives

$$\pi_1' = \pi_{11}p_1' + \pi_{12}p_2' - q_{11} = 0,$$

$$\pi_2' = \pi_{12}p_1' + \pi_{22}p_2' - q_{12} = 0.$$

Differentiating (40) and (41), then putting $p_1 = p_2 = 1, t = 0$, and substituting for the π_{ij} and q_{ij}, enables us to solve the last two equations, obtaining

$$p_1' = -\tfrac{1}{147}, \qquad p_2' = -\tfrac{39}{147}.$$

In competition the reduction of price through taxation could be displayed by linear demand and cost functions. This is not true for monopoly. For in this case all the g_{ij} and p_{ijk} would vanish, and (39) would reduce to $p_{ij} = \tfrac{1}{2}\pi_{ij}$. When this is substituted in (35), an elementary property of determinants applied in connection with the definition of σ_{ij} gives $p_i' = \tfrac{1}{2}\delta_{i1}$. Hence p_i' cannot be negative. Thus, the second derivatives of the demand and cost functions play an essential part.

Consider now such a case as that of two stores in the same city, identical in all respects, symmetrically placed with regard to customers, and selling just one commodity. Suppose them controlled by a single owner. Will a tax levied upon the quantity sold in one of the stores cause him to raise or to lower his prices?

This is essentially the case we have been discussing of a monopolist selling two articles. The demand wil be correlated, since an increase in price at one of the stores will send some customers to the other, and we shall have $p_{12} < 0$. Even if there are other competing stores selling the same commodity, the case is essentially little different. The feature whose consequences we will now examine is the *symmetry*.

The demand functions will, on account of the symmetry with respect to the market, be such that $f_1(q_1, q_2) = f_2(q_2, q_1)$, while the cost functions will, on account of the identical nature of the two stores, have the property that $g_1(q_1, q_2) = g_2(q_2, q_1)$. For any such symmetry it must be that $\pi_{11} = \pi_{22}$, and consequently $\sigma_{11} = \sigma_{22}$; also $p_{22} = p_{11}$. The relation $\sigma_{11}\sigma_{22} - \sigma_{12}^2 > 0$ gives, in this case,

$$\sigma_{11}^2 - \sigma_{12}^2 > 0.$$

From (36) and (37) we deduced that $\sigma_{12} > 0$ if Edgeworth's phenomenon occurs. If we assume this, we may multiply (37) by σ_{12} and (36) by $-\sigma_{11}$, which is always positive, and add. Putting $p_{22} = p_{11}$, this gives:

$$-p_{11}(\sigma_{11}^2 - \sigma_{12}^2) < 0.$$

Since $p_{11} < 0$, this is an absurdity. Hence Edgeworth's phenomenon cannot arise in a perfectly symmetrical situation.

But even with symmetry it is possible that the price at the untaxed store may be lowered as a result of the tax, though the price at the taxed store is increased. It is easy to construct such cases; for instance, with zero costs,

$$q_1 = a - 2(p_1 - c) + (p_2 - d) + \tfrac{1}{4}(p_1 - c)^2$$
$$- \tfrac{1}{2}(p_1 - c)(p_2 - d) - \tfrac{1}{4}(p_2 - d)^2,$$
$$q_2 = b + (p_1 - c) - 2(p_2 - d) - \tfrac{1}{4}(p_1 - c)^2$$
$$- \tfrac{1}{2}(p_1 - c)(p_2 - d) + \tfrac{1}{4}(p_2 - d)^2.$$

Here a, b, c, d may be any positive quantities satisfying

$$a = 2c - d, \qquad b = -c + 2d.$$

9. Three or More Commodities

That the prices of three or more commodities may be lowered as a result of a tax imposed upon one of them follows from the fact that this is possible for two commodities, because a continuous array of degrees of similarity can be conceived, in such a way that several commodities merge gradually into one. The discussion in the last section is largely applicable to any number of commodities. However a few new features enter.

For n commodities, the equations analogous to (39) are $n(n + 1)/2$ in number. The number of independent unknowns p_{ijk}, the order of the subscripts being immaterial, is

$$\frac{n(n + 1)(n + 2)}{6},$$

and is thus considerably in excess of the number of equations. That these equations are actually consistent will appear when the matrix is shown to contain a non-vanishing determinant of order $n(n + 1)/2$. If we arrange equations and unknowns in the order of the first index and then in the order of the second, such a determinant will stand at the left of the matrix. Its principal diagonal will consist entirely of 1's, with zeros below.

The analysis of the last section might have been carried out with the prices instead of the quantities as the independent variables. This involves introducing quantities ρ_{ij} conjugate to $\partial^2 \pi / \partial p_i \, \partial p_j$. The ρ_{ij} are the coefficients in a negative definite quadratic form. In terms of these quantities,

$$p_i' = \sum_j \rho_{ij} q_{1j}.$$

From this a proposition of some interest follows: *If all the prices are lowered by a tax, the taxed commodity must compete with at least one of the others.* If this were not true, all the derivatives $q_{11}, q_{12}, \ldots, q_{1n}$, would be negative or zero. Since each p_i' is to be negative, we should then have

$$\sum p_i' q_{1i} = \sum \sum \rho_{ij} q_{1i} q_{1j} > 0.$$

But the second member of this equation is a quadratic form which must be negative definite, so that there is a contradiction.

Articles sold by the monopolist which complement that taxed, or which, so far as first-order terms in the expansion go, are neither complementary nor competitive with it, may also decrease in price as a result of the tax, provided the monopolist sells at least three different commodities. Examples may easily be constructed to show this.

10. SUMMARY

Edgeworth's examples proving that a monopolist may find his most profitable course in lowering the prices of two articles which he sells as a result of a tax which he must pay on one of them we have simplified, at the same time setting forth the conditions upon which the phenomenon depends. These conditions include asymmetry of cost or of demand for the two commodities, as well as the analytical conditions of Section 8. In Section 2 a simple example with discontinuous demand is given to illustrate the phenomenon.

Though Edgeworth attributed the phenomenon to conditions of monopoly, it has been shown in Section 7 that a tax on sellers of two commodities may result in both prices being lowered even under free competition. Necessary, though not sufficient, conditions for this are that the commodities compete both in consumption and in production. Many agricultural products satisfy these conditions.

In monopoly it is necessary that the commodities "compete" for customers, in the sense that each may, to some extent, replace the other. If a monopolist sells three or more commodities whose prices will be lowered when a tax is imposed upon one of them, the taxed commodity must be a possible or partial substitute for at least one of the others, but not necessarily for all.

Edgeworth's phenomenon has generally been considered very improbable in the actual economic world. The basis for this belief seems to be that demand functions constructed at random usually do not display it, together with an elementary proof that for a commodity unrelated to any other a tax leads to an increase in price. But this simplified case is not realized in the complex economic system which exists; and the construction of hypothetical systems at random depends upon irrelevant features of the convenience and mathematical equipment of the student rather than upon objective truth. The algorisms given in Sections 7 and 8 for the construction of examples make clear that there are as many degrees of freedom in situations displaying the phenomenon as in those which do not, both in competition and in monopoly.

In order to make sure that all the relevant conditions are examined in coming to the conclusion that there is nothing improbable in the phenomenon, a more intensive study is made in this paper than had previously been undertaken of the conditions which demand and supply functions may reasonably be expected to satisfy. This study yields results which have a direct bearing upon statistical studies of demand and supply, and leads to suggestions regarding the direction which these studies may well take. In the study of groups of commodities, considerations enter which are of a different nature from those involved in the study of single commodities.

The theory of probability is used in Section 4 to prove that the demand or the supply function for a single commodity is of the nature of an ogive and that such functions may often be expected to have the form of the integral of the normal probability distribution. This opens up interesting possibilities of using such mathematical theories as that of semi-invariants in a combined attack by statistical and by technological studies upon the urgent and difficult problem of finding definite demand and supply functions.

Columbia University   HAROLD HOTELLING

Note on Edgeworth's Taxation Phenomenon and Professor Garver's Additional Condition on Demand Functions

In a recent paper[1] I derived for demand functions for n commodities conditions which, I concluded, assure the satisfaction of all further conditions which may reasonably be applied to *all* sets of demand functions. Letting $p_1, p_2, \ldots, p_n$, and $q_1, q_2, \ldots, q_n$ be respectively the prices and quantities, the conditions are:

$$\frac{\partial p_i}{\partial q_j} = \frac{\partial p_j}{\partial q_i}, \frac{\partial p_i}{\partial q_i} < 0, \qquad \frac{\partial(p_i, p_j)}{\partial(q_i, q_j)} > 0, \qquad \frac{\partial(p_i, p_j, p_k)}{\partial(q_i, q_j, q_k)} < 0, \ldots .$$

These are equivalent to

$$\frac{\partial q_i}{\partial p_j} = \frac{\partial q_j}{\partial p_i}, \frac{\partial q_i}{\partial p_i} < 0, \qquad \frac{\partial(q_i, q_j)}{\partial(p_i, p_j)} > 0, \qquad \frac{\partial(q_i, q_j, q_k)}{\partial(p_i, p_j, p_k)} < 0, \ldots,$$

and were known, at least in part, by Edgeworth. Similar conditions in which the inequality signs are all $>$ apply for supply functions.

One important feature of these conditions is that they are invariants. Thus if we change the units of measure, or even replace the commodities by bundles containing mixtures of them in varying proportions, these conditions, if satisfied originally, will be satisfied by the new units or the new bundles. It seems reasonable that any set of conditions of universal application, as distinguished from the characteristics of particular commodities, should be invariants, just as in physics it is accepted that a law of nature should not depend on particular coordinates but should be capable of invariantive expression. Certainly it is obvious that general economic principles should at least be independent of the question whether one commodity is measured in miles and another in ounces.

Professor Garver's new condition, which we may write in the same notation,

$$\frac{\partial q_1}{\partial p_1} + \frac{\partial q_1}{\partial p_2} < 0,$$

[1] "Edgeworth's Taxation Paradox and the Nature of Demand and Supply Functions," *Journal of Political Economy*, XL (1932), p. 590.

is not an invariant, even under changes in the unit of measure of one of the commodities. It is therefore not of universal application to pairs of commodities, though individual cases can be found in which it seems reasonable. But the seeming reasonableness may arise in an illusory way from the accidental use of units having the same name for measuring different commodities. The fact that butter and margarine are both sold by the pound does not mean that a pound of margarine is a substitute for exactly a pound of butter. For consumers who will pay as much for two pounds of margarine as for one pound of butter, equal small increases in price will increase the consumption of butter, in violation of Professor Garver's condition. A somewhat similar point is dealt with on pp. 595–6* of the paper cited above.

Since the new condition is applicable only in particular cases, its non-satisfaction in a hypothetical example does not show the example unrealistic. The examples which Edgeworth gave, and the additional examples in my paper cited above, cannot be attacked on this ground.

The necessity of a cost function to produce Edgeworth's phenomenon does not follow from the new condition, for this is satisfied by the example of my equations (40). It arises rather from arbitrarily taking the demand functions to be linear.

In spite of these qualifying considerations, Professor Garver's paper is of real value in pointing out a condition which may well be expected in numerous cases. He has also performed a service for those concerned with the statistical evaluation of demand functions in emphasizing that theoretical conditions exist with which empirical results may be compared.

Columbia University

* *Editor's note*: Pages 106–7 of the reprinted article in this volume.

Demand Functions with Limited Budgets*

I

In the study of related commodities it is important to consider the properties that may reasonably be attributed to the demand functions. Taking these functions in the form

$$p_i = f_i(q_1, q_2, \ldots, q_n) \qquad (i = 1, 2, \ldots, n), \tag{1.1}$$

where p_i and q_i denote respectively the price and the quantity of the ith of n commodities, and putting

$$p_{ij} = \frac{\partial p_i}{\partial q_j}, \tag{1.2}$$

the conditions

$$p_{ij} = p_{ji}, \tag{1.3}$$

together with the inequalities on the Jacobian determinants

$$\frac{\partial p_i}{\partial q_i} < 0, \quad \frac{\partial(p_i, p_j)}{\partial(q_i, q_j)} > 0, \quad \frac{\partial(p_i, p_j, p_k)}{\partial(q_i, q_j, q_k)} < 0, \ldots, \quad \text{and}$$

$$\frac{\partial q_i}{\partial p_i} < 0, \quad \frac{\partial(q_i, q_j)}{\partial(p_i, p_j)} > 0, \quad \frac{\partial(q_i, q_j, q_k)}{\partial(p_i, p_j, p_k)} < 0, \ldots, \tag{1.4}$$

which generalize the condition that a demand curve shall decline, have been deduced[1] on the assumption that the buyers are entrepreneurs, or others

---

* Presented at the Berkeley meeting of the Econometric Society and the American Statistical Association, 20 June, 1934.

[1] Harold Hotelling, "Edgeworth's Taxation Paradox and the Nature of Demand and Supply Functions," *Journal of Political Economy*, XL (1932), 571–616, especially p. 590. An interesting statistical study involving a test of the conditions (1.3) and the equivalent conditions

$$\frac{\partial q_j}{\partial p_i} = \frac{\partial q_i}{\partial p_j}$$

is contained in a paper by Henry Schultz: "Interrelations of Demand," *Journal of Political Economy*, XLI (August, 1933). Among the four commodities, barley, corn, hay, and oats, six adjusting their purchases for maximum profit, in such a way that their purchases could be increased without effective limit if they should lead to sufficiently increased profits. Similar conditions were deduced for supply functions. In case a buyer's budget is limited, it was pointed out that his demand functions will not, in general, satisfy the conditions (1.3), but will satisfy

$$p_i(p_{jk} - p_{kj}) + p_j(p_{ki} - p_{ik}) + p_k(p_{ij} - p_{ji}) = 0. \qquad (1.5)$$

We now inquire what inequalities are to replace (1.4) in case of a limited budget. The results obtained are applicable not only to the purchases of individuals, where we are dealing with commodities for actual consumption, and forming a substantial part or all of their expenditures, but likewise to departments of corporations or governments, which often have their budgets fixed in advance, with little opportunity for variation on account of price changes within a budget period.

Results similar to those of the present paper may be obtained for supply functions of a certain character. Thus, a subsistence farmer, who must get a certain cash return to meet taxes and necessary money expenditures, might be treated as having a negative budget, which must be filled by some selection of cash crops which the farmer will seek to produce with a minimum of disutility. Conditions on such supply functions may be derived from the results that we shall obtain by changing the sign of m, the amount of the budget. Negative quantities may be regarded as sales, positive quantities as purchases. The prices, however, will never be negative.

II

As a preliminary, we shall retrace the former argument, with more detailed attention to the process of passing from the demand functions of individuals to the total demand functions, which can be examined statistically. We suppose there are N competing buyers of n commodities, and use, in this section only, a system of subscripts in which Greek letters denote individual buyers, while Latin letters correspond to the commodities. Suppose that the αth buyer, by purchasing quantities $q_{1\alpha}, q_{2\alpha}, \ldots, q_{n\alpha}$ of the several commodities can produce and sell goods that will bring him a gross money income u_α. This gross income we write as a function of the $q_{i\alpha}$:

$$u_\alpha = u_\alpha(q_{1\alpha}, \ldots, q_{n\alpha}).$$

---

equations of each of these types are tested. In most cases the differences between the left and right members as calculated statistically appear to be well within the discrepancies to be expected on the basis of the standard errors, though one of the six differences (hay-oats) is too great to be ascribed to chance alone. Schultz' results as a whole seem to confirm the applicability to these commodities of the integrability conditions. It must be remembered that even though a single discrepancy may be judged significant when it exceeds double its standard error, still, among six, it is quite probable that one will fall beyond this limit. The standard errors of the differences tested are somewhat difficult to determine, since the partial derivatives are taken as regression coefficients found from separate least-square solutions, between which the residuals, and therefore the solutions, are correlated. No exact theory exists for dealing with this situation with full statistical efficiency.

The net income of this buyer is then

$$\pi_\alpha = u_\alpha - p_1 q_{1\alpha} - p_2 q_{2\alpha} - \cdots - p_n q_{n\alpha}. \tag{2.1}$$

This is to be made a maximum. The buyer is assumed to regard the n prices as fixed by the market, independent of his purchases. Hence,

$$\frac{\partial \pi_\alpha}{\partial q_{i\alpha}} = \frac{\partial u_\alpha}{\partial q_{i\alpha}} - p_i = 0. \tag{2.2}$$

Defining $f_{i\alpha}$ as

$$f_{i\alpha} = f_{i\alpha}(q_{1\alpha}, q_{2\alpha}, \ldots, q_{n\alpha}) = \frac{\partial u_\alpha}{\partial q_{i\alpha}}, \tag{2.3}$$

we thus have

$$p_i = f_{i\alpha}(q_{1\alpha}, \ldots, q_{n\alpha}) \qquad (i = 1, 2, \ldots, n) \tag{2.4}$$

as the demand functions for this buyer. Evidently, on account of (2.3), they satisfy the integrability conditions

$$\frac{\partial f_{i\alpha}}{\partial q_{j\alpha}} = \frac{\partial f_{j\alpha}}{\partial q_{i\alpha}}, \tag{2.5}$$

which may be written more briefly in the form

$$\frac{\partial p_i}{\partial q_{j\alpha}} = \frac{\partial p_j}{\partial q_{i\alpha}}. \tag{2.6}$$

Also, the n equations (2.4) may be solved to give the quantities that this buyer will purchase at given prices, in the form

$$q_{i\alpha} = F_{i\alpha}(p_1, p_2, \ldots, p_n). \tag{2.7}$$

These last functions satisfy integrability conditions similar to (2.6). Indeed, differentiating p_i with respect to p_k,

$$\sum_j \frac{\partial p_i}{\partial q_{j\alpha}} \frac{\partial q_{j\alpha}}{\partial p_k} = \delta_{ik}, \tag{2.8}$$

where $\delta_{ik} = 0$ if $i \neq k$, but equals unity if $i = k$. Holding k fixed but letting i vary from 1 to n, we have in (2.8) n equations in the n unknowns

$$\frac{\partial q_{1\alpha}}{\partial p_k}, \frac{\partial q_{2\alpha}}{\partial p_k}, \ldots, \frac{\partial q_{n\alpha}}{\partial p_k}.$$

Putting H_α for the determinant in which the element in the ith row and jth column is $\partial p_i / \partial q_{j\alpha}$, we have as the solution,

$$\frac{\partial q_{j\alpha}}{\partial p_k} = \frac{\text{cofactor of } \dfrac{\partial p_k}{\partial q_{j\alpha}} \text{ in } H_\alpha}{H_\alpha}, \tag{2.9}$$

a result familiar in implicit function theory. But from (2.6) it follows that H_α

is a symmetrical determinant, whence, from (2.9),

$$\frac{\partial q_{j\alpha}}{\partial p_k} = \frac{\partial q_{k\alpha}}{\partial p_j}. \tag{2.10}$$

Now letting

$$q_i = F_i(p_1, \ldots, p_n) \qquad (i = 1, 2, \ldots, n) \tag{2.11}$$

be the total demand functions for the whole aggregate of buyers, we have

$$q_j = \sum_\alpha q_{j\alpha}. \tag{2.12}$$

Hence, summing (2.10) with respect to α,

$$\frac{\partial q_j}{\partial p_k} = \frac{\partial q_k}{\partial p_j}. \tag{2.13}$$

Furthermore, if we solve (2.11) for the p_i, thus writing the total demand functions in the form (1.1), these functions must satisfy

$$\frac{\partial p_i}{\partial q_j} = \frac{\partial p_j}{\partial q_i}, \tag{2.14}$$

in accordance with the same argument used in deducing (2.10) from (2.6). These are the conditions (1.3).

When the foregoing first-order conditions are satisfied, the buyer's profit π_α will be a maximum if the second derivatives of π_α, which are the same as those of u_α, are the coefficients of a negative definite quadratic form. With the notation used in (2.6), this means that

$$\sum_i \sum_j \frac{\partial p_i}{\partial q_{j\alpha}} x_i x_j \tag{2.15}$$

must be negative definite. Since the conjugate of a negative definite form is negative definite, we have from (2.9) that

$$\sum \sum \frac{\partial q_{j\alpha}}{\partial p_i} x_i x_j \tag{2.16}$$

is likewise negative for all values of $x_1, x_2, \ldots, x_n$ which do not all vanish. Upon summing (2.16) with respect to α and using (2.12), we find that

$$\sum \sum \frac{\partial q_j}{\partial p_i} x_i x_j \tag{2.17}$$

is likewise negative definite. This conclusion is equivalent to the conditions on the total demand functions (2.11),

$$\frac{\partial q_i}{\partial p_i} < 0, \qquad \frac{\partial(q_i, q_j)}{\partial(p_i, p_j)} > 0, \qquad \frac{\partial(q_i, q_j, q_k)}{\partial(p_i, p_j, p_k)} < 0, \ldots. \tag{2.18}$$

From the negative definite character of (2.17) follows that of its conjugate, from which follow the conditions (1.4).

For supply functions a similar argument leads to positive definite forms, with results which differ from the foregoing formulae only in that all the inequality signs are $>$.

III

Turning now to the case of a buyer with a limited budget m, which will be assumed positive, we refer to the concept of a utility function φ which this person desires to make a maximum. This function φ of the quantities purchased is of the nature of a rank index, distinguishing one "indifference surface" from another, and need not be regarded as uniquely defined. Indeed, any other function ψ of the quantities, provided only that ψ is an increasing function of φ with a sufficient number of continuous derivatives, will be equally satisfactory. From this it will follow that our results must have an invariantive character, in that they may be expressed in terms either of φ or of ψ, and each of these statements must be a logical consequence of the other.

We shall hereafter omit the additional subscript α used in the last section to distinguish among individuals, and denote the quantities a particular person buys merely by $q_1, \ldots, q_n$.

Subject to the condition

$$\sum_{i=1}^{n} p_i q_i = m, \tag{3.1}$$

the function

$$\varphi = \varphi(q_1, q_2, \ldots, q_n) \tag{3.2}$$

is to be a maximum. The first-order conditions for this are

$$\varphi_i = \lambda p_i \qquad (i = 1, 2, \ldots, n) \tag{3.3}$$

where $\varphi_i = \partial\varphi/\partial q_i$, and the Lagrange multiplier λ is the "marginal utility of money", and is essentially positive, since both p_i and φ_i are positive, the latter because φ is assumed to be an increasing function of the quantities.

The demand functions are found by eliminating λ between (3.1) and (3.3), in the form

$$p_i = \frac{m\varphi_i}{\sum_j \varphi_j q_j}. \tag{3.4}$$

Denoting differentiation with respect to q_j by an added subscript j, we have from (3.3)

$$\varphi_{ij} = \lambda p_{ij} + \lambda_j p_i. \tag{3.5}$$

Since $\varphi_{ij} = \varphi_{ji}$, the condition (1.5) may be obtained from (3.5) by interchanging i and j and subtracting, multiplying by p_k, permuting i, j, and k cyclically, and adding the three equations thus obtained.

The second-order condition[2] for a maximum of φ subject to the constraint

---

[2] Harris Hancock: *Theory of Maxima and Minima* (1917), pp. 115–116.

(3.1) is that each root of the equation in σ,

$$
\begin{vmatrix}
\varphi_{11} - \sigma & \varphi_{12} & \cdots & \varphi_{1n} & p_1 \\
\varphi_{21} & \varphi_{22} - \sigma & \cdots & \varphi_{2n} & p_2 \\
\hdotsfor{5} \\
\varphi_{n1} & \varphi_{n2} & \cdots & \varphi_{nn} - \sigma & p_n \\
p_1 & p_2 & \cdots & p_n & 0
\end{vmatrix} = 0, \tag{3.6}
$$

shall be negative. The roots are known all to be real, on account of the symmetry of the determinant and the reality of the φ_{ij} and p_i.

If a function of n variables possesses a maximum for a certain set of values of these variables, whether or not there are subsidiary conditions, and if some of these variables are held fixed at their maximizing values while the others vary, the function must under this new type of variation still be a maximum for the same values as before. It follows that the second-order conditions applicable to the whole set of variables are likewise applicable to every subset of them, so long as any degree of freedom remains. In particular, rows other than the last, and the like-ordered columns, may be deleted from (3.6), and the resulting equations must have all their roots negative. At least three rows must, however, remain in the reduced determinant in order that it may actually involve σ and thus express a condition.

In (3.6) substitute the values (3.5) of the second derivatives. Then from the jth column subtract the product of λ_j by the last column. Divide each column but the last by λ, and multiply the last row by λ. Upon putting

$$\rho = \sigma/\lambda,$$

so that ρ has the same sign as λ, this gives the result that all the roots of

$$
\begin{vmatrix}
p_{11} - \rho & p_{12} & \cdots & p_{1n} & p_1 \\
p_{21} & p_{22} - \rho & \cdots & p_{2n} & p_2 \\
\hdotsfor{5} \\
p_{n1} & p_{n2} & \cdots & p_{nn} - \rho & p_n \\
p_1 & p_2 & \cdots & p_n & 0
\end{vmatrix} = 0, \tag{3.7}
$$

must be negative. In like manner, all roots of all symmetrical determinantal equations formed from (3.7) by deleting rows and columns other than the last must be negative. Thus we have obtained the conditions upon the first derivatives of the demand functions (3.4) which correspond to those of the last section. These conditions may however be put in a simpler form.

Let

$$
\Delta_{12} = \begin{vmatrix}
p_{11} & p_{12} & p_1 \\
p_{21} & p_{22} & p_2 \\
p_1 & p_2 & 0
\end{vmatrix}, \quad
\Delta_{123} = \begin{vmatrix}
p_{11} & p_{12} & p_{13} & p_1 \\
p_{21} & p_{22} & p_{23} & p_2 \\
p_{31} & p_{32} & p_{33} & p_3 \\
p_1 & p_2 & p_3 & 0
\end{vmatrix}, \tag{3.8}
$$

and likewise for other combinations of subscripts. Upon deleting from (3.7) all rows and columns other than the first two and last, and expanding in powers of ρ, the resulting equation is

$$(p_1^2 + p_2^2)\rho + \Delta_{12} = 0.$$

Since the root is to be negative, Δ_{12} must be positive, i.e., the individual demand functions for two related commodities must satisfy the inequality:

$$\frac{\partial p_1}{\partial q_1} p_2^2 - \left(\frac{\partial p_1}{\partial q_2} + \frac{\partial p_2}{\partial q_1}\right) p_1 p_2 + \frac{\partial p_2}{\partial q_2} p_1^2 < 0. \tag{3.9}$$

Expanding the four-rowed principal minor of (3.7) which corresponds to the commodities numbered 1, 2 and 3, we have

$$-(p_1^2 + p_2^2 + p_3^2)\rho^2 - (\Delta_{12} + \Delta_{23} + \Delta_{13})\rho + \Delta_{123} = 0,$$

of which both roots are to be negative. This gives $\Delta_{123} < 0$, and no further conditions, since we already know that Δ_{12}, and by the same reasoning Δ_{23} and Δ_{13}, are positive. Proceeding in this way, and noting that Descartes' rule of signs requires that for negative roots all the coefficients in an equation having real roots must be of like sign, we find that our second-order conditions are equivalent to the requirement that *each of the determinants* $\Delta_{ijk} \ldots$ *must have the sign of* $(-1)^s$, *where s is the number of commodities represented in the determinant.* Thus:

$$(-1)^s \Delta_{i_1 i_2 \ldots i_s} > 0. \tag{3.10}$$

Any of the inequalities with which we deal may in special cases be replaced by equalities, in case a maximum occurs at a point at which the representing surface has contact of specially high order with its tangent plane. However, in empirical functions, this is infinitely improbable, and we need give little attention to the possibility.

The indifference loci are $(n-1)$-dimensional hypersurfaces in an n-dimensional space in which the quantities taken by the individual are Cartesian coordinates. He has no preference for one point over another on a fixed indifference locus, but desires to move from one to another of these loci in the direction of increasing quantities. However, he is constrained to remain on the hyperplane whose equation is (3.1). The collection of goods which he can buy with his limited budget, at the prices that have been fixed beyond his control and that determine the orientation of his hyperplane, will give him greatest satisfaction if he chooses their quantities equal to the coordinates of a point of tangency of the hyperplane with an indifference locus. The coordinates of this point of tangency satisfy the equations (3.4). The inequalities on the determinants $\Delta_{ijk} \ldots$ are geometrically equivalent to the requirement that, at the point of tangency chosen, the hypersurface shall be concave toward the directions of increasing quantities. The actual quantities being supposed positive, this means that the hypersurface must be convex to the origin at any point that can actually represent the purchases of an individual who makes the best of his budget. In particular, for two commodities, we have a system of indifference curves, one of which will be tangent to the straight line determined by the prices and the amount of money to be spent. For maximum utility this indifference curve must, at the point of contact, bend away from the straight line in the direction of increasing quantities.

If indifference curves for purchases be thought of as possessing a wavy character, convex to the origin in some regions and concave in others, we are forced to the conclusion that it is only the portions convex to the origin that can be regarded as possessing any importance, since the others are essentially unobservable. They can be detected only by the discontinuities that may occur in demand with variation of price-ratios, leading to an abrupt jumping of a point of tangency across a chasm when the straight line is rotated. But, while such discontinuities may reveal the existence of chasms, they can never measure their depth. The concave portions of the indifference curves and their many-dimensional generalizations, if they exist, must forever remain in unmeasurable obscurity.

IV

The ordinary conception of a demand curve involves a *ceteris paribus* assumption that is peculiarly objectionable, in that there is not only the observed fact that other things do not remain equal during the variation of the price and quantity of a particular commodity, but that it is logically impossible that they do so. Indeed, for n related commodities (and all commodities are related to others), we may have the demand situation expressed in the form

$$p_i = f_i(q_1, q_2, \ldots, q_n) \qquad (i = 1, 2, \ldots, n) \qquad (4.1)$$

or in the alternative form, found by solving these equations,

$$q_j = F_j(p_1, p_2, \ldots, p_n). \qquad (4.2)$$

In tracing the relation between p_1 and q_1, the graph of the first equation in (4.1) when $q_2, \ldots, q_n$ are held constant will be different from and inconsistent with the equally valid "demand curve" obtained from (4.2) by holding $p_2, \ldots, p_n$ constant. There is a real difference between fixing the prices and fixing the quantities of the commodities related to the first, and these two fixations are not simultaneously possible.

The sets of expressions (4.1) and (4.2) may be regarded as alternative standard forms of the demand functions for n commodities related to each other but not to others. Besides these standard forms, however, there are many other possible ways of expressing the same demand relations. Indeed, any set of n independent equations among the p's and q's would suffice to define these relations. Certain of these alternative forms of demand functions will now be used to express the inequalities of the last section in another way which lends itself to the combination, in some circumstances, of these inequalities upon individual demand functions into inequalities upon the total demand functions.

Instead of solving the whole set of equations (4.1) for the q's, let us select only s of them, denoting their indices by i_1, i_2, ..., i_s, and solve these for q_{i_1}, q_{i_2}, ..., q_{i_s}. Let the remaining $n - s$ indices be denoted by i_{s+1}, ..., i_n. The solution may be written:

$$q_\beta = F_{\beta \cdot i_{s+1} \ldots i_n}(p_{i_1}, \ldots, p_{i_s}, q_{i_{s+1}}, \ldots, q_{i_n}) \qquad (\beta = i_1, \ldots, i_s). \qquad (4.3)$$

We now proceed to determine the derivatives of these functions which we denote by

$$q_{\beta\gamma \cdot i_{s+1} \ldots i_n} = \frac{\partial F_{\beta \cdot i_{s+1} \ldots i_n}}{\partial p_\gamma} \qquad (\beta, \gamma = i_1, \ldots, i_s) \qquad (4.4)$$

in terms of the derivatives p_{ij} of (4.1). The derivative (4.4) differs from what we have previously called $\partial q_\beta / \partial p_\gamma$ in that the quantities instead of the prices of $n - s$ of the commodities are held constant during the differentiation.

When the expressions (4.3) are substituted for $q_{i_1}, \ldots, q_{i_s}$ in (4.1), the resulting equations are identities. Differentiating the ith of these identities with respect to p_γ, we have, since,

$$\frac{\partial f_i}{\partial q_\beta} = p_{i\beta}, \qquad \text{while} \qquad \frac{\partial p_i}{\partial p_\gamma} = \delta_{i\gamma}, \qquad (4.5)$$

where $\delta_{i\gamma}$ is unity if $i = \gamma$ but is zero if $i \neq \gamma$,

$$\sum_{\beta=i_1}^{i_s} p_{i\beta} q_{\beta\gamma \cdot i_{s+1} \ldots i_n} = \delta_{i\gamma}. \qquad (4.6)$$

Keeping γ fixed as one of the numbers $i_1, \ldots, i_s$, and letting i vary through this set of numbers, we have in (4.6) a set of s linear equations in the unknowns (4.4). Denote the determinant of the coefficients by

$$J_{i_1 i_2 \ldots i_s} = \begin{vmatrix} p_{i_1 i_1} & p_{i_1 i_2} & \cdots & p_{i_1 i_s} \\ \cdots\cdots\cdots\cdots\cdots \\ \cdots\cdots\cdots\cdots\cdots \\ p_{i_s i_1} & \cdots\cdots\cdots & p_{i_s i_s} \end{vmatrix}. \qquad (4.7)$$

Then the solution of (4.6) is:

$$q_{\beta\gamma \cdot i_{s+1} \ldots i_n} = \frac{\text{cofactor of } p_{\gamma\beta} \text{ in } J_{i_1 \ldots i_s}}{J_{i_1 \ldots i_s}}. \qquad (4.8)$$

Now expand an s-rowed determinant of the type (3.8) with reference to its last row and last column. With the help of (4.8) and (3.10) the result may be written:

$$(-1)^{s+1}\Delta_{i_1 \ldots i_s} = J_{i_1 \ldots i_s} \sum\sum p_\beta p_\gamma q_{\beta\gamma \cdot i_{s+1} \ldots i_n} < 0. \qquad (4.9)$$

In this expression the sums are with respect to β and γ, which range over values $i_i, i_2, \ldots, i_s$.

If, for example, we take $s = 2$, we have from (4.9)

$$-\Delta_{12} = J_{12}[p_1^2 q_{11 \cdot 34 \ldots n} + p_1 p_2(q_{12 \cdot 3 \ldots n} + q_{21 \cdot 3 \ldots n}) + p_2^2 q_{22 \cdot 3 \ldots n}] < 0,$$

where

$$J_{12} = p_{11}p_{22} - p_{12}p_{21}.$$

V

Now consider the additional hypothesis that, for each buyer in the market, the determinant (4.7) has the sign of $(-1)^s$. In this case (4.9) takes the form

$$(-1)^s \sum \sum p_\beta p_\gamma q_{\beta\gamma \cdot i_{s+1} \ldots i_n} < 0, \qquad (5.1)$$

a relation valid for each buyer. Upon summing for all the buyers, since the p's are the same for all, and since the derivative $q_{\beta\gamma \cdot i_{s+1} \ldots i_n}$ of the total quantity bought is the sum of such derivatives of quantities bought by individuals, we find that *the inequalities* (5.1) *hold when applied to the market as a whole*, provided the hypothesis regarding the signs of the determinants is satisfied.

In this case, if we assume also that the determinant $J_{i_1 \ldots i_s}$ pertaining to the whole market also has the sign of $(-1)^s$, we can pass back from (5.1) to (4.9), and thence to (3.10). Thus, from the hypotheses mentioned, it follows that s-rowed determinants such as (3.8) will have the same sign for the total demand functions as for the individual demand functions.

That this assumption regarding the signs of the determinants (4.7) may reasonably be expected to be satisfied in an extended class of cases is indicated by the argument of II, since the conclusion (1.4) is equivalent to the assumption here made. The fact that (1.4) consists of inequalities rather than equations is enough to show that its validity extends beyond the scope of the assumptions of II from which (1.4) was derived. Thus there is considerable reason to expect that the inequalities (3.10) and (4.9) can be applied to total demand functions. This inference is strengthened by the further observation that it is not necessary to the conclusion that *all* the buyers should have their determinants (4.7) of the signs assumed, but only a preponderance of them in the weighted average formed by adding together the second members of (4.9).

One further argument adds weight to our conclusion that the inequalities we have derived for individual demand functions can with considerable probability be supposed to hold for total demand functions. For the case $s = n$, that is, the case involving all the commodities among which the budget is to be apportioned, we can actually prove that the determinant (4.7) has, for each individual, the sign assumed for it.

VI

Differentiating the individual demand function (3.4), which may be written

$$p_i = m \frac{\phi_i}{\sum \phi_k q_k},$$

with respect to q_j, we obtain

$$p_{ij} = \frac{\partial p_i}{\partial q_j} = m \frac{\phi_{ij} \sum \phi_k q_k - \phi_i (\phi_j + \sum \phi_{jk} q_k)}{(\sum \phi_k q_k)^2}. \qquad (6.1)$$

From (3.1) and (3.3),

$$\sum \phi_k q_k = \lambda m. \tag{6.2}$$

Substituting this in (6.1) gives

$$p_{ij} = \frac{\lambda m \phi_{ij} - \phi_i(\phi_j + \sum \phi_{jk} q_k)}{\lambda^2 m}. \tag{6.3}$$

Substituting this in the n-rowed determinant of the p_{ij} and multiplying each row and column by $\lambda^2 m$, we have

$$\lambda^{2n} m^n J_{12\ldots n} = \begin{vmatrix} \lambda m \phi_{11} - \phi_1(\phi_1 + \sum \phi_{1k} q_k) & \cdots & \lambda m \phi_{1n} - \phi_1(\phi_n + \sum \phi_{nk} q_k) \\ \cdots\cdots\cdots\cdots\cdots\cdots\cdots\cdots\cdots\cdots\cdots\cdots\cdots\cdots\cdots\cdots\cdots \\ \cdots\cdots\cdots\cdots\cdots\cdots\cdots\cdots\cdots\cdots\cdots\cdots\cdots\cdots\cdots\cdots\cdots \\ \lambda m \phi_{n1} - \phi_n(\phi_1 + \sum \phi_{1k} q_k) & \cdots & \lambda m \phi_{nn} - \phi_n(\phi_n + \sum \phi_{nk} q_k) \end{vmatrix}. \tag{6.4}$$

Let us border this determinant with zeros on the right, with unity in the lower right-hand corner, and with

$$\phi_j + \sum \phi_{jk} q_k$$

at the foot of the jth column ($j = 1, 2, \ldots, n$). In the bordered determinant, add to the ith row the product of ϕ_i by the last row ($i = 1, \ldots, n$). Thus we obtain from (6.4)

$$\lambda^{2n} m^n J_{1\ldots n} = \begin{vmatrix} \lambda m \phi_{11} & \cdots & \lambda m \phi_{1n} & \phi_1 \\ \cdots\cdots\cdots\cdots\cdots\cdots\cdots\cdots\cdots\cdots\cdots\cdots \\ \lambda m \phi_{n1} & \cdots & \lambda m \phi_{nn} & \phi_n \\ \phi_1 + \sum \phi_{1k} q_k & \cdots & \phi_n + \sum \phi_{nk} q_k & 1 \end{vmatrix}.$$

Multiply the last column of this determinant by λm and divide each of the first n rows by the same quantity. Then subtract from the last row the product of q_k by the kth row ($k = 1, \ldots, n$). The element in the lower right corner vanishes in accordance with (6.2), and we have

$$\lambda^{n+1} m J_{1\ldots n} = \begin{vmatrix} \phi_{11} & \cdots & \phi_{1n} & \phi_1 \\ \cdots\cdots\cdots\cdots\cdots\cdots \\ \phi_{n1} & \cdots & \phi_{nn} & \phi_n \\ \phi_1 & \cdots & \phi_n & 0 \end{vmatrix}. \tag{6.5}$$

Let us now revert to the equation (3.6), whose roots must all be negative. In accordance with Descartes' rule of signs and the reality of the roots, this requires that the coefficients of all powers of σ be of the same sign. In particular, the coefficient of the highest power of σ must have the same sign as the term independent of σ. But the former is

$$(-1)^n (p_1^2 + p_2^2 + \cdots + p_n^2),$$

while the term independent of σ, obtained by putting $\sigma = 0$ in (3.6), is simply λ^2 times the determinant in (6.5), by (3.3). This establishes the desired conclusion,

$$(-1)^n J_{12\ldots n} > 0,$$

since $m > 0$.

It can now be shown that *all* the Jacobian determinants (4.7) of order two or more, calculated from individual demand functions, and not simply those involving all the commodities in the system, have the signs assumed at the beginning of the fifth section above. This is not an independent assumption, but can be proved by the reasoning used in the last section. It is only necessary to replace m by the amount of money spent on the subgroup of commodities considered, and to change slightly the meaning of λ, so that it is a function of these rather than of all the quantities. The legitimacy of this follows from the remark in the paragraph following (3.6). This greatly strengthens the probability that the inequalities proved for individual buyers can all be applied to total demand functions, and definitely proves that (5.1) can thus be applied.

Columbia University

Curtailing Production Is Anti-Social

The success of the government's recovery program, or of any national economic program, must be judged, not in terms of price levels, but in terms of the quantities of physical goods and services which are put into the hands of consumers. With this is to be considered the effect of the program on the distribution of wealth among different classes. But the chief thing needed is to increase physical production. In this respect much that is being done at Washington is definitely in the wrong direction. The attempts to increase the prices and curtail the production of oil, agricultural products, and other commodities are anti-social.

Instead of decreasing production, the government might well take measures to increase production, by intervening to lower certain important prices which have not fallen as much as the general level during the depression, and in some cases have actually risen. Most notable among these prices, as barriers to prosperity, are railroad and utility rates. It is suggested that the government operate the railroads and utilities, charging far lower prices for their services than now prevail, and paying overhead costs by means of income, inheritance, and land taxes. This would bring about a tremendous increase in traffic and in business generally, just as the abolition of tolls on roads and bridges has been followed by increases in traffic. The extent of the increase in prosperity resulting from such a move has been investigated mathematically by the author, using the theory of line integrals and integrability conditions.

The idea that all the overhead of such an enterprise as a railroad ought to be paid out of operating revenues is examined mathematically, and found to rest on a fallacy.

The success of any national economic program must be judged eventually by two criteria. One of these is its effect on the distribution of the total of wealth and income among the various elements of the population, that is, on the manner in which the national dividend is divided. The other is the effect on the total of what is to be divided. In the pulling and hauling of politics, the first of these considerations is usually uppermost, since to each class in the community it seems that its main chance for betterment through governmental action is by getting the government to help this class to take something away from other classes, or to keep what it has in the face of aggression from other classes. Of course this is not the language used in public appeals by the embattled farmers, railroads, veterans, naval interests, oil companies, debtors, and other special groups. Each industry, in demanding higher prices and special favors for itself, is careful to avoid suggesting any other particular group which is to bear the burden of the increased prosperity of the industry in question, and the attempt is frequently made to convey the idea that prosperity for this particular industry is likely to increase the prosperity of the country as a whole. Put into practice, this leads to the familiar attempts to lift the economic organization by its bootstraps through such devices as the tariff, and other oblique subsidies to particular groups.

The loud cries of special groups for help at the general expense were intensified during the depression, and served effectively to divert attention from the simple principle that in order to consume it is first necessary for society to produce. By a curious inversion, the terrible lack of physical goods, which recently has brought millions to distress and thousands to actual death by starvation, is ascribed to "overproduction."

This popular view that we have too much of everything, and that therefore measures need to be taken to destroy stocks and reduce production is embodied in several leading features of the administration's program. Besides slaughtering and burying pigs, and paying farmers to plow under their cotton and curtail their cereal crops, the government assisted the oil companies in their successful attempt to curtail the flow of oil and the output of refined products, with the consequence that motorists must drive fewer miles and pay more for their gasoline. Not only has the reduction in output resulted in much loss of employment for labor in the oil fields and refineries, and the closing of many service stations which formerly prospered along the highways, as well as diminishing the public revenues from gasoline taxes, but since less gasoline ineluctably implies less driving, production control measures cannot but diminish the use and pleasure derived from motor vehicles.

In contrast with the earnest efforts to obtain a redistribution of wealth by means of limitation of output and by inflation, a redistribution which might be accomplished more efficiently by paying direct subsidies to the favored classes and taxing the rest, the problem of producing something to distribute receives very little public notice. Yet by certain reorganizations it is possible to bring about such gains in the efficiency of the economic mechanism as to insure to the entire population a level of prosperity exceeding any in history, with a flow of goods accompanied by increased leisure, and moreover a diminished danger of the cyclical fluctuations which from time to time threaten to wreck our whole social structure. Such a program should bring the country into a highly prosperous condition within a very brief time.

From the standpoint of national prosperity, rather than of the distribution of wealth among classes and individuals, the key to increased general satisfaction lies in the full and efficient utilization of productive equipment. All about us we see idle factories, idle railroad equipment, and idle man-power.

Even in our times of "prosperity," there is a tremendous under-utilization of equipment and resources which could with advantage be fully utilized. This is not all or chiefly a result of ignorance or stupidity on the part of the owners of the equipment; it is more largely an inescapable consequence of the necessity of continuing in business and paying dividends. This goes back to a fundamental fault in the economic position of those industries in which the element of overhead cost is large. It has often been said that "every tub must stand on its own bottom", and the inference has been drawn that every industry should sell its products at prices sufficiently high to cover, not merely the out-of-pocket direct cost of production, but also the interest on the investment in productive equipment, and other overhead costs. This theory is at the bottom of a very large share of our difficulties.

As a matter of fact, the efficient way to operate a bridge—and the same applies to a railroad or a factory, if we neglect the small cost of an additional unit of product or transportation—is to make it free to the public, so long at least as the use of its does not increase to a state of overcrowding. A free bridge costs no more to construct than a toll bridge, and costs less to operate; but society, which must pay the cost in some way or other, gets far more benefit from the bridge if it is free, since in this case it will be more used. Charging a toll, however small, causes some people to waste time and money in going around by longer but cheaper ways, and prevents others from crossing. The higher the toll, the greater is the damage done in this way; to a first approximation, for small tolls, the damage is proportional to the square of the toll rate. There is no such damage if the bridge is paid for by income, inheritance, and land taxes, or for example by a tax on the real estate benefited, with exemption of new improvements from taxation, so as not to interfere with the use of the land. The distribution of wealth among members of the community is affected by the mode of payment adopted for the bridge, but not the total wealth, except that it is diminished by bridge tolls and other similar forms of excise. This is such plain common sense that toll bridges have now largely disappeared from civilized communities. But New York City bridge and tunnels across the Hudson are still operated on a toll basis, due to the pressure of real estate interests anxious to shift the tax burden to wayfarers and the possibility of collecting considerable sums from persons who do not vote in the city.

These are the pertinent considerations if the bridge is already in existence, or its construction definitely decided. But if we examine the general question of the circumstances in which bridges ought to be built, a further inefficiency is disclosed in the scheme of paying for bridges out of tolls. For society, it is often beneficial to build a bridge under conditions in which no scheme of tolls could possibly make it profitable from the standpoint of a private owner.

The argument about bridges applies equally to railroads, except that in the latter case there is some slight additional cost resulting from an extra passenger or an extra shipment of freight. My weight is such that when I ride on the train, more coal has to be burned in the locomotive, and I wear down the station platform by walking across it. What is more serious, I may help to overcrowd the train, diminishing the comfort of other travelers and helping to create a situation in which additional trains should be run, but often are not. The trivial nature of the extra costs of marginal use of the railroads is evidently realized by the railroad managements themselves, for it is implied in the amazingly complex rate structures they build up in the attempt to squeeze the last possible bit of revenue from freight and passenger traffic. If in a rational economic system the railroads were operated for the benefit of the people as a whole, it is plain that if people were to be induced by low rates to travel in one season rather than another, the season selected should be one in which travel would otherwise be light, leaving the cars nearly empty, and not a season in which they are normally overcrowded. Actually, our railroads run trains about the country in winter with few passengers, while crowding multitudes of sweating travellers into their cars in summer. The rates are made high in winter, lower in summer, on the ground that the summer demand is more elastic than that of the winter travelers, who are usually on business rather than pleasure, and thus decide the question of a trip with less sensitiveness to the cost.

The General Welfare in Relation to Problems of Taxation and of Railway and Utility Rates*

In this paper we shall bring down to date in revised form an argument due essentially to the engineer Jules Dupuit, to the effect that the optimum of the general welfare corresponds to the sale of everything at marginal cost. This means that toll bridges, which have recently been reintroduced around New York, are inefficient reversions; that all taxes on commodities, including sales taxes, are more objectionable than taxes on incomes, inheritances, and the site value of land; and that the latter taxes might well be applied to cover the fixed costs of electric power plants, waterworks, railroads, and other industries in which the fixed costs are large, so as to reduce to the level of marginal cost the prices charged for the services and products of these industries. The common assumption, so often accepted uncritically as a basis of arguments on important public questions, that "every tub must stand on its own bottom," and that therefore the products of every industry must be sold at prices so high as to cover not only marginal costs but also all the fixed costs, including interest on irrevocable and often hypothetical investments, will thus be seen to be inconsistent with the maximum of social efficiency. A method of measuring the loss of satisfactions resulting from the current scheme of pricing, a loss which appears to be extremely large, will emerge from the analysis. It will appear also that the inefficient plan of requiring that all costs, including fixed overhead, of an industry shall be paid out of the prices of its products is responsible for an important part of the instability which leads to cyclical fluctuations and unemployment of labor and other resources.

A railway rate is of essentially the same nature as a tax. Authorized and enforced by the government, it shares with taxes a considerable degree of arbitrariness. Rate differentials have, like protective tariffs and other taxes, been used for purposes other than to raise revenue. Indeed, the difference between rail freight rates between the same points, according as the commodity is or is not moving in international transport, has been used in effect to nullify the protective tariff. While it has not generally been perceived that the

* Presented at the meeting of the Econometric Society at Atlantic City, 28 December 1937, by the retiring president.

problems of taxation and those of railway rate making are closely connected, so that two independent bodies of economic literature have grown up, nevertheless the underlying unity is such that the considerations applicable to taxation are very nearly identical with those involved in proper rate making. This essential unity extends itself also to other rates, such as those charged by electric, gas, and water concerns, and to the prices of the products of all industries having large fixed costs independent of the volume of output.

I. THE CLASSICAL ARGUMENT

Dupuit's work of 1844 and the following years[1] laid the foundation for the use of the diagram of Figure 1 by Marshall and other economists. A rising supply curve SB is used, and is sometimes regarded as coinciding with the marginal-cost curve. Such a coincidence would arise if there were free competition among producers, in the sense that each would regard the price as fixed beyond his control, and adjust his production so as to obtain maximum net profits. This condition is approximated, for example, in most agriculture. DB is a declining demand curve. The buyers are presumed to compete freely with each other. The actual quantity and price are the co-ordinates of the intersection B. Then it is supposed that a tax t per unit is imposed upon the sellers. Since this is a uniform increment to marginal cost, the marginal-cost curve SB is lifted bodily to the new position RL, at height $t = SR = NL$ above its former position.

Three conclusions have been derived with the help of this figure, all of which must be reviewed to take account of the interrelations of the particular commodity in question with others. One of these arguments has almost universally been accepted, but must be rejected when account is taken of related commodities. A second has been accepted, and is actually true. The third has been condemned and attacked by a long line of prominent economists, but in the light of the more thorough analysis made possible by modern mathematical methods must now in its essence be accepted. The first is the proposition that since the point L of intersection of the demand curve with the supply curve RL is higher by GL, a fraction of the tax rate NL, than the intersection B with the tax-free curve SB, therefore the price is increased as a result of the tax, by an amount less than the tax. That this conclusion is not necessarily true when account is taken of related commodities I have shown in an earlier paper.[2] The second proposition—whose conclusion remains

[1] Collected and reprinted with comments by Mario di Bernardi and Luigi Einaudi. "De l'Utilité et de sa Mesure," *La Riforma Sociale*, Turin, 1932.

[2] "Edgeworth's Taxation Paradox and the Nature of Demand and Supply Functions," *Journal of Political Economy*, Vol. 40, 1932, pp. 577–616. Edgeworth had discovered, and maintained against the opposition of leading economists, that a monopolist controlling two products may after the imposition of a tax on one of them find it profitable to reduce both prices, besides paying the tax. However he regarded this as a "mere curiosum," unlikely in fact to occur, and peculiar to monopoly. But it is shown in the paper cited that the phenomenon is also possible with free competition, and is quite likely to occur in many cases, either under monopoly or under competition.

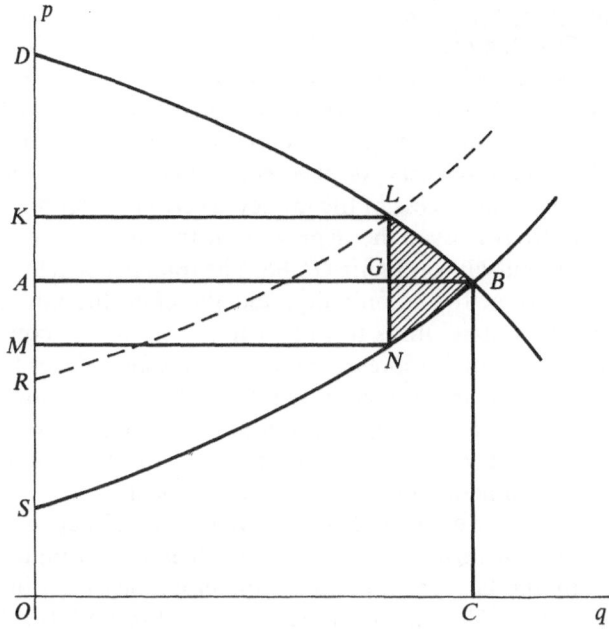

FIGURE 1

valid under certain plausible assumptions[3]—is that, since L is to the left of B, the quantity of the taxed commodity will diminish. With this diminution is associated an approximately measurable net social loss.

The third argument is based on Dupuit's, and is of primary concern here. Dupuit sought a criterion of the value to society of roads, canals, bridges, and waterworks. He pointed out the weakness of calling the value of a thing only what is paid for it, since many users would if necessary pay more than they actually do pay. The total benefit he measured by the aggregate of the maximum prices that would be paid for the individual small units of the commodity (a term used here to include services, e.g., of canals) corresponding to the costs of alternatives to the various uses. If $p = f(q)$ is the cost of the best alternative to the use of an additional small unit of the commodity when q units are already used, then, if q_0 units are used altogether,

$$\int_0^{q_0} f(q)\, dq \tag{1}$$

is the total benefit, which Dupuit called *utilité*, resulting from the existence of the canal or other such facility making possible the commodity (service) in

---

[3] On p. 600 of the paper just cited the conclusion is reached that it is reasonable to regard the matrix of the quantities h_{ij} as negative definite. From this and equation (19) of that page it follows that a positive increment in the tax t_j on the jth commodity causes a negative increment in the quantity of this commodity. *Editor's note*: See p. 110 of the current reprint.

question. Since $p = f(q)$ is the ordinate of the demand curve DB in Figure 1, this total benefit is the total area under the arc DB. To obtain what is now called the *consumers' surplus* we must subtract the amount paid by consumers, namely the product of the price by the quantity, represented by the rectangle $OCBA$. Thus the consumers' surplus is represented by the curvilinear triangle ABD. There is also a *producers' surplus* represented by the lower curvilinear triangle SBA; this is the excess of the money received by producers (the area of the rectangle $OCBA$) over the aggregate of the marginal costs, which is represented by the curvilinear figure $OCBS$. The total net benefit, representing the value to society of the commodity, and therefore the maximum worth spending from the public funds to obtain it, is the sum of consumers' and producers' surpluses, and is represented by the large curvilinear triangle SBD. It is the difference between the integral (1) of the demand function and the integral between the same limits of the marginal-cost function.

Imposition of the tax, by raising the price to the level of KL, appears to reduce the consumers' surplus to the curvilinear area KLD. The new producers' surplus is the area RLK, which equals SNM. There is also a benefit on account of the government revenue, which is the product of the new quantity MN by the tax rate NL, and is therefore measured by the area of the rectangle $MNLK$. The sum of these three benefits is $SNLD$. It falls short of the original sum of producers' and consumers' surpluses by the shaded triangular area NBL.

This shaded area represents the net social loss due to the tax, and was discovered by Dupuit. If the tax is small enough, the arcs BL and NB may be treated as straight lines, and the area of the triangle is, to a sufficient approximation, half the product of the base NL by the altitude GB. Since GB is the decrement in the quantity produced and consumed because of the tax, and NL is the tax rate, we may say that the net loss resulting from the tax is half the product of the tax rate by the decrement in quantity. But since the decrement in quantity is, for small taxes, proportional to the tax rate, it then follows that the net loss is proportional to the *square* of the tax rate. This fact also was remarked upon by Dupuit.

This remarkable conclusion has frequently been ignored in discussions in which it should, if correct, be the controlling consideration. The open attacks upon it seem all to be based on an excessive emphasis on the shortcomings of consumers' and producers' surpluses as measures of benefits. These objections are four in number: (1) Since the demand curve for a necessity might for very small quantities rise to infinity, the integral under the curve might also be infinite. This difficulty can be avoided by measuring from some selected value of q greater than zero. Since in the foregoing argument it is only *differences* in the values of the surpluses that are essentially involved, it is not necessary to assign exact values. The situation is the same as in the physical theory of the potential, which involves an arbitrary additive constant and so may be measured from any convenient point, since only its differences are important. (2) Pleasure is essentially nonmeasurable and so, it is said, cannot be represented by consumers' surplus or any other numerical magnitude. We shall meet this objection by establishing a generalized form of Dupuit's conclusion on the basis of a ranking only, without measurement, of satisfactions, in the way represented graphically by indifference curves. The same analysis will dispose also of the objections (3) that the consumers' surpluses arising from different commodities are not independent and cannot be added to each other, and (4) that the surpluses of different persons cannot be added.

In connection with the last two points, it will be observed that if we have a set of n related commodities whose demand functions are

$$p_i = f_i(q_1, q_2, \ldots, q_n) \qquad (i = 1, 2, \ldots, n),$$

then the natural generalization of the integral representing total benefit, of which consumers' surplus is a part, is the line integral

$$\int (f_1 \, dq_1 + f_2 \, dq_2 + \cdots + f_n \, dq_n), \qquad (2)$$

taken from an arbitrary set of values of the q's to a set corresponding to the actual quantities consumed. The net benefit is obtained by subtracting from (2) a similar line integral in which the demand functions $f_1, f_2, \ldots, f_n$ are replaced by the marginal-cost functions

$$g_i(q_1, q_2, \ldots, q_n) \qquad (i = 1, 2, \ldots, n).$$

If we put

$$h_i = f_i - g_i,$$

the total net benefit is then measured by the line integral

$$w = \int \sum h_i \, dq_i. \qquad (3)$$

Such indeterminacy as exists in this measure of benefit is only that which arises with variation of the value of the integral when the path of integration between the same end points is varied. The condition that all these paths of integration shall give the same value is that the integrability conditions

$$\frac{\partial h_i}{\partial q_j} = \frac{\partial h_j}{\partial q_i}$$

be satisfied. In the paper on "Edgeworth's Taxation Paradox" already referred to, and more explicitly in a later note,[4] I have shown that there is a good reason to expect these integrability conditions to be satisfied, at least to a close approximation, in an extensive class of cases. If they are satisfied, the surpluses arising from different commodities, and also the surpluses belonging to different persons, may be added to give a meaningful measure of social value.

---

[4] "Demand Functions with Limited Budgets," *Econometrica*, Vol. 3, 1935, pp. 66–78. A different proof is given by Henry Schultz in the *Journal of Political Economy*, Vol. 41, 1933, p. 478.

This breaks down if the variations under consideration are too large a part of the total economy of the person or the society in question; but for moderately small variations, with a stable price level and stable conditions associated with commodities not in the group, the line integral w seems to be a very satisfactory measure of benefits. It is invariant under changes in units of measure of the various commodities, and also under a more general type of change of our way of specifying the commodities, such as replacing "bread" and "beef" by two different kinds of "sandwiches." For these reasons the total of all values of w seems to be the best measure of welfare that can be obtained without considering the proportions in which the total of purchasing power is subdivided among individuals, or the general level of money incomes. The change in w that will result from a proposed new public enterprise, such as building a bridge, may fairly be set against the cost of the bridge to decide whether the enterprise should be undertaken. It is certainly a better criterion of social value than the aggregate $\sum p_i q_i$ of tolls that can be collected on various classes of traffic, as Dupuit pointed out for the case of a single commodity or service. The actual calculation of w in such a case would be a matter of estimation of vehicular and pedestrian traffic originating and terminating in particular zones, with a comparison of distances by alternative routes in each case, and an evaluation of the savings in each class of movement. Determination whether to build the bridge by calculation merely of the revenue $\sum p_i q_i$ obtainable from tolls is always too conservative a criterion. Such public works will frequently be of great social value even though there is no possible system of charging for their services that will meet the cost.

II. The Fundamental Theorem

But without depending in any way on consumers' or producers' surpluses, even in the form of these line integrals, we shall establish a generalization of Dupuit's result. We take our stand on the firm ground of a system of preferences expressible by a function

$$\Phi = \Phi(q_1, q_2, \ldots, q_n)$$

of the quantities $q_1, q_2, \ldots, q_n$ of goods or services consumed by an individual per unit of time. If the function Φ, Pareto's *ophélimité*, has the same value for one set of q's as for another, then the one combination of quantities is as satisfactory to the individual in question as the other. For two commodities, Φ is constant along each of a set of "indifference curves"; and likewise for n commodities, we may think of a system of hypersurfaces of which one passes through each point of a space of n dimensions, whose Cartesian co-ordinates are the quantities of the various goods. These hypersurfaces we shall refer to as *indifference loci*.

It is to be emphasized that the indifference loci, unlike measures of pleasure, are objective and capable of empirical determination. One interesting experimental attack on this problem was made by L. L. Thurstone, who by means of questionnaires succeeded in mapping out in a tentative manner the indifference loci of a group of girls for hats, shoes, and coats.[5] Quite a different method, involving the study of actual family budgets, also appears promising.[6] The function Φ, on the other hand, is not completely determinable from observations alone, unless we are prepared to make some additional postulate about independence of commodities, as was done by Irving Fisher in defining utility,[7] and by Ragnar Frisch.[8] The present argument does not depend on any such assumption, and therefore allows the replacement of Φ by an arbitrary increasing function Ψ of Φ, such as sinh Φ, or $\Phi + \Phi^3$. The statements we shall make about Φ will apply equally to every such function Ψ. Negative values of the q's are the quantities of labor, or of goods or services, produced by the individual. It is with the understanding that this kind of indeterminacy exists that we shall sometimes refer to Φ and Ψ as utility functions.

Consider now a state in which income and inheritance taxes are used to pay for the construction of bridges, roads, railroads, waterworks, electric power plants, and like facilities, together with other fixed costs of industry; and in which the facilities may be used, or the products of industry consumed, by anyone upon payment of the additional net cost occasioned by the particular use or consumption involved in each case. This additional net cost, or marginal cost, will include the cost of the additional labor and other resources required for the particular item of service or product involved, beyond what would be required without the production of that particular item. Where facilities are not adequate to meet all demands, they are made so either by enlargement, or by checking the demand through inclusion in the price of a rental charge for the facilities, adjusted so as to equate demand to supply. Such a rental cost, of which the site rental of land is an example, is an additional source of revenue to the state; it must not be confused with carrying charges on invested capital, or with overhead cost. Some such charge is necessary to discriminate economically among would-be users of the facilities. Another example is that of water in a dry country; if demand exceeds supply, and no enlargement of supply is possible, a charge must be made for the water sufficient to reduce the demand to the supply. Such a charge is an element of marginal cost as here defined.

The individual retains, after payment of taxes, a money income m. At prices $p_1, p_2, \ldots, p_n$ determined in the foregoing manner, he can buy or sell such quantities $q_1, q_2, \ldots, q_n$ as he pleases, subject to the condition that

$$\sum p_i q_i = m. \tag{4}$$

---

[5] "The Indifference Function," *Journal of Social Psychology*, Vol. 2, 1931, pp. 139–167, esp. pp. 151 ff.

[6] R. G. D. Allen and A. L. Bowley, *Family Expenditure*, London, 1935.

[7] *Mathematical Investigations in the Theory of Value and Prices*, New Haven, 1892.

[8] *New Methods of Measuring Marginal Utility*, Tübingen, 1932. Dr. Frisch also considered the possibility of substitute commodities in his *Confluence Analysis*, and in collaboration with Dr. F. V. Waugh made an attempt to handle this situation statistically.

The combination he chooses will be such as to make his indifference function Φ a maximum, subject to the condition (4). We may put aside as infinitely improbable—having probability zero, though not impossible—the contingency that two different sets of values of the q's satisfying (4) will give the same degree of satisfaction. We therefore have that, if $q_1, \ldots, q_n$ are the quantities chosen under these conditions, and if $q'_1, \ldots, q'_n$ are any other set of quantities satisfying (4), so that

$$\sum p_i q'_i = m, \tag{5}$$

then

$$\Phi = \Phi(q_1, \ldots, q_n) > \Phi(q'_1, \ldots, q'_n) = \Phi + \delta\Phi,$$

say. Hence, putting $q'_i = q_i + \delta q_i$ in (5) and subtracting (4), we find that any set of values of $\delta q_1, \ldots, \delta q_n$ satisfying

$$\sum p_i\, \delta q_i = 0, \tag{6}$$

and not all zero, must have the property that

$$\delta\Phi = \Phi(q_1 + \delta q_1, \ldots, q_n + \delta q_n) - \Phi(q_1, \ldots, q_n) < 0. \tag{7}$$

Let us now consider an alteration of the system by the imposition of excise taxes and reduction of income taxes. Some of the taxes may be negative; that is, they may be bounties or subsidies to particular industries; or, instead of being called taxes, they may be called tolls, or charges for services or the use of facilities over and above marginal cost. There ensues a redistribution of production and consumption. Let p_i, q_i, and m be replaced respectively by

$$p'_i = p_i + \delta p_i, \qquad q'_i = q_i + \delta q_i, \qquad m' = m + \delta m, \tag{8}$$

where the various increments δp_i, δq_i are not constrained to be either positive or negative; some may have one sign and some the other. The yield of the new excise taxes will be the sum, over all individuals, of the quantity which for the particular individual we are considering is $\sum q'_i\, \delta p_i$. (We use the sign $\sum$ to denote summation over all commodities, including services.) Since this person's income tax is reduced by δm, the net increment of government revenue

$$\delta r = \sum q'_i\, \delta p_i - \delta m \tag{9}$$

may be imputed to him, in the sense that summation of δr over all persons gives the total increment of government revenue.[9] We neglect changes in administrative costs and the like.

[9] A friendly critic writes "It is not clear to me why δp_i should be the exact per-unit revenue of the state from an excise tax which raises the price by δp_i from its old level.... I should expect (referring to Figure 1) an increase in price of GL, and a revenue to the state of NL." The answer to this is that the summation of δr over all persons includes the sellers as well as the buyers, and that the government revenue per unit of the commodity is derived in part from each—though it must be understood that the contribution of either or both may be negative. In the classical case represented by Figure 1, the buyers' δp is the height GL, while the sellers' is NG in magnitude and is negative. Since q' is positive for the buyer and negative for the seller, the product $q'\, \delta p$ is in each case positive. The aggregate of these positive terms is the total tax revenue from the commodity.

The individual's budgetary limitation now takes the form $\sum p_i' q_i' = m'$, which may also be written

$$\sum (p_i + \delta p_i)(q_i + \delta q_i) = m + \delta m. \tag{10}$$

Subtracting the budget equation (4) corresponding to the former system and using (8) we find that

$$\delta m = \sum q_i' \, \delta p_i + \sum p_i \, \delta q_i. \tag{11}$$

Substituting this in (9) we find that

$$\delta r = -\sum p_i \, \delta q_i. \tag{12}$$

Suppose that, to avoid disturbing the existing distribution of wealth, the excise taxes paid by each individual (in the sense of incidence just defined; not in the sense of handing over the money to the government in person) are exactly offset by the decrement in his income tax. Then $\delta r = 0$. From (12) it then follows that (6) is satisfied. Except in the highly improbable case of all the δq's coming out exactly zero, it would then follow from (7) that this man's new state is worse than his old. The change from income to excise taxes has resulted in a net loss of satisfactions. Conversely, if we start from a system of excise taxes, or any system in which sales are not at marginal cost, this argument shows that there is a possible distribution of personal income taxes such that everyone will be better satisfied to change to the system of income taxes with sales at marginal cost. The problem of the distribution of wealth and income among persons or classes is not involved in this proposition.

This argument may be expressed in geometrical language as follows: Let $q_1, \ldots, q_n$ be Cartesian coordinates in a space of n dimensions. Through each point of this space passes a hypersurface whose equation may be written $\Phi(q_1, \ldots, q_n) = $ constant. The individual's satisfaction is enhanced by moving from one to another of these hypersurfaces if the value of the constant on the right side of the equation is thereby increased; this will usually correspond to moving in a direction along which some or all of the q's increase. The point representing the individual's combination of goods is however constrained in the first instance to lie in the hyperplane whose equation is (4). In this equation the p's and m are to be regarded as constant coefficients, while the q's vary over the hyperplane. A certain point Q on this hyperplane will be selected, corresponding to the maximum taken by the function Φ, subject to the limitation (4). If the functions involved are analytic, Q will be the point of tangency of the hyperplane with one of the "indifference loci." The change in the tax system means that the individual must find a point Q' in the new hyperplane whose equation is $\sum p_i' q_i = m'$. If we denote the coordinates of Q' by $q_1', \ldots, q_n'$, we have, upon substituting them in the equation of this new hyperplane, $\sum p_i' q_i' = m'$. If the changes in prices and m are such as to leave the government revenue unchanged, (12) must vanish; that is,

$$\sum p_i q_i' = \sum p_i q_i.$$

Since $\sum p_i q_i = m$, this shows that $\sum p_i q_i' = m$; that is, that Q' lies on the same hyperplane to which Q was confined in the first place. But since Q was chosen among all the points on this hyperplane as the one lying on the outermost possible indifference locus, for which Φ is a maximum, and since we are putting aside the infinitely improbable case of there being other points on the hyperplane having this maximizing property, it follows that Q' must lie on some other indifference locus, and that this will correspond to a lesser degree of satisfaction.

The fundamental theorem thus established is that *if a person must pay a certain sum of money in taxes, his satisfaction will be greater if the levy is made directly on him as a fixed amount than if it is made through a system of excise taxes which he can to some extent avoid by rearranging his production and consumption.* In the latter case, the excise taxes must be at rates sufficiently high to yield the required revenue *after* the person's rearrangement of his budget. The redistribution of his production and consumption then represents a loss to him without any corresponding gain to the treasury. This conclusion is not new. What we have done is to establish it in a rigorous manner free from the fallacious methods of reasoning about one commodity at a time which have led to false conclusions in other associated discussions.

The conclusion that a fixed levy such as an income or land tax is better for an individual than a system of excise taxes may be extended to the whole aggregate of individuals. In making this extension it is necessary to neglect certain interactions among the individuals that may be called "social" in character, and are separate and distinct from the interactions through the economic mechanisms of price and exchange. An example of such "social" interactions is the case of the drunkard who, after adjusting his consumption of whisky to what he considers his own maximum of satisfaction, beats his wife, and makes his automobile a public menace on the highway. The restrictive taxation and regulation of alcoholic liquors and certain other commodities do not fall under the purview of our theorems because of these social interactions which are not economic in the strict sense. With this qualification, and neglecting also certain possibilities whose total probability is zero, we have:

If government revenue is produced by any system of excise taxes, there exists a possible distribution of personal levies among the individuals of the community such that the abolition of the excise taxes and their replacement by these levies will yield the same revenue while leaving each person in a state more satisfactory to himself than before.

It is in the sense of this theorem that we shall in later sections speak of "the maximum of total satisfactions" or "the maximum of general welfare" or "the maximum national dividend" requiring as a necessary, though not sufficient, condition that the sale of goods shall be without additions to price in the nature of excise taxes. These looser expressions are in common use, and are convenient; when used in this paper, they refer to the proposition above, which depends only on rank ordering of satisfactions; there is no connotation of adding utility functions of different persons.

The inefficiency of an economic system in which there are excise taxes or bounties, or in which overhead or other charges are paid by excesses of price over marginal cost, admits of an approximate measure when the deviations from the optimum system described above are not great, if, as is customary in this and other kinds of applied mathematics, we assume continuity of the indifference function and its derivatives. Putting for brevity

$$\Phi_i = \frac{\partial \Phi}{\partial q_i}, \qquad \Phi_{ij} = \frac{\partial^2 \Phi}{\partial q_i\, \partial q_j},$$

we observe that the maximum of Φ, subject to the budget equation (4), requires that

$$\Phi_i = \lambda p_i \qquad (i = 1, 2, \ldots, n), \tag{13}$$

where the Lagrange multiplier λ is the marginal utility of money. Differentiating this equation gives

$$\Phi_{ij} = \lambda \frac{\partial p_i}{\partial q_j} + p_i \frac{\partial \lambda}{\partial q_j}. \tag{14}$$

Expanding the change in the utility or indifference function we obtain, with the help of (13), (12), and (14),

$$\delta \Phi = \sum \Phi_i\, \delta q_i + \frac{1}{2} \sum \sum \Phi_{ij}\, \delta q_i\, \delta q_j + \cdots$$

$$= -\lambda \delta r + \frac{1}{2} \lambda \sum \sum \frac{\partial p_i}{\partial q_j}\, \delta q_i\, \delta q_j - \frac{1}{2} \delta r \sum \frac{\partial \lambda}{\partial q_j}\, \delta q_j + \cdots, \tag{15}$$

where the terms omitted are of third and higher order, and are therefore on our assumptions negligible. Their omission corresponds to Dupuit's deliberate neglect of curvilinearity of the sides of the shaded triangle in Figure 1. With accuracy of this order we have further,

$$\delta p_i = \sum_j \frac{\partial p_i}{\partial q_j}\, \delta q_j, \qquad \delta \lambda = \sum_j \frac{\partial \lambda}{\partial q_j}\, \delta q_j.$$

Upon substituting for these expressions, (15) reduces to

$$\delta \Phi = -\lambda \delta r + \tfrac{1}{2} \lambda \sum \delta p_i\, \delta q_i - \tfrac{1}{2} \delta r\, \delta \lambda + \cdots. \tag{16}$$

If the readjustment from the original state of selling only at marginal cost, with income taxes to pay overhead, is such as to leave $\delta r = 0$ as above, (16) reduces to

$$\delta \Phi = \tfrac{1}{2} \lambda \sum \delta p_i\, \delta q_i + \cdots, \tag{17}$$

where the terms omitted are of higher order.

As another possibility we may consider a substitution of excise for income tax so arranged as to leave this person's degree of satisfaction unchanged. Upon putting $\delta \Phi = 0$ in (16) and solving for δr we have, apart from terms of higher order,

$$\delta r = \tfrac{1}{2} \sum \delta p_i\, \delta q_i + \cdots. \tag{18}$$

This is the net loss to the state in terms of money, so far as this one individual is concerned. The net loss in terms of satisfactions is merely the product of (18) by the marginal utility of money λ, that is, (17), if we neglect terms of higher order than those written. The total net loss of state revenue resulting from abandonment of the system of charging only marginal costs, and uncompensated by any gain to any individual, is the sum of (18) over all individuals. If the prices are the same for all, this sum is of exactly the same form as the right-hand member of (18), with δq_i now denoting the increment (positive or negative) of the total quantity of the ith commodity.

The approximate net loss

$$\tfrac{1}{2} \sum \delta p_i \, \delta q_i \tag{19}$$

may be regarded as the sum of the areas of the shaded triangles in the older graphic demonstration. It should however be remembered that the readjustment of prices caused by excise taxes is not necessarily in the direction formerly supposed, that some of the quantities and some of the prices may increase and some decrease, and that some of the terms of the foregoing sum may be positive and some negative. But the aggregate of all these varying terms is seen by the foregoing argument to represent a dead loss, and never a gain, as a result of a change from income to excise taxes, or away from a system of sales at marginal cost. Any inaccuracy of the measure (19) is of only the same order as the error involved in replacing the short arcs LB and NB in Figure 1 by straight segments, and can never affect the sign.

It is remarkable, and may appear paradoxical, that without assuming any particular measure of utility or any means of comparison of one person's utility with another's, we have been able to arrive at (19) as a valid approximation measuring in money a total loss of satisfactions to many persons. That the result depends only on the conception of ranking, without measurement, of satisfactions by each person is readily apparent from the foregoing demonstration; or we may for any person replace Φ by another function Ψ as an index of the same system of ranks among satisfactions. If we do this in such a way that the derivatives are continuous, we shall have $\Psi = F(\Phi)$, where F is an increasing function with continuous derivatives. Upon writing the expressions for the first and second derivatives of Ψ in terms of those of F and Φ it may be seen that the foregoing formulae involving Φ are necessary and sufficient conditions for the truth of the same equations with Ψ written in place of Φ. The result (18) is independent of which system of indicating ranks is used. The fundamental fact here is that *arbitrary* analytic transformations, even of very complicated functional forms, always induce *homogeneous linear* transformations of differentials.

Not only the approximation (19) but also the whole expression indicated by (18) are absolutely invariant under all analytic transformations of the utility functions of all the persons involved. These expressions depend only on the demand and supply functions, which are capable of operational determination. They represent simply the money cost to the state of the inefficiency of the system of excise taxation, when this is arranged in such a way as to leave unchanged the satisfactions derived from his private income by each person.

The arguments based on Figure 1 have been repeated with various degrees of hesitation, or rediscovered independently, by numerous writers including Jevons, Fisher, Colson, Marshall, and Taussig. Marshall considered variations of the figure involving downward-sloping cost curves and multiple solutions, and was led to the proposal (less definite than that embodied in the criterion established by our theorem) that incomes and increasing-cost industries be taxed to subsidize decreasing-cost ones. He observed the difficulty of defining demand curves and consumers' surplus in view of the interdependence of demand for various commodities. These difficulties are indeed such that it now seems better to stop talking about demand *curves*, and to substitute demand *functions*, which will in general involve many variables, and are not susceptible of graphic representation in two or three dimensions. Marshall was one of those misled by Figure 1 into thinking that a tax of so much per unit imposed on producers of a commodity leads necessarily to an increase of price by something less than the tax.

Though the marginal-cost curve in Figure 1 slopes upward, no such assumption is involved in the present argument. It is perfectly possible that an industry may be operated by the state under conditions of diminishing marginal cost. The criterion for a small increase in production is still that its cost shall not exceed what buyers are willing to pay for it; that is, the general welfare is promoted by offering it for sale at its marginal cost. It may be that demand will grow as prices decline until marginal cost is pushed to a very low level, far below the average cost of all the units produced. In such a case the higher cost of the first units produced is of the same character as fixed costs, and is best carried by the public treasury without attempting to assess it against the users of the particular commodity as such. Our argument likewise makes no exception of cases in which more than one equilibrium is possible. Where there are multiple solutions we have that sales at marginal cost are a necessary, though not a sufficient, condition for the optimum of general welfare.

The confusion between marginal and average cost must be avoided. This confusion enters into many of the arguments for laissez-faire policies. It is frequently associated with the calm assumption, as a self-evident axiom, that the whole costs of every enterprise must be paid out of the prices of its products. This fallacious assumption appears, for example, in recent writings on government ownership of railroads. It has become so ingrained by endless repetition that it is not even stated in connection with many of the arguments it underlies.

III. Tax Systems Minimizing Dead Loss

The magnitude of the dead loss varies greatly according to the objects taxed. While graphic arguments are of suggestive value only, it may be observed from Figure 1 that the ratio of the dead loss NBL to the revenue $MNLK$ depends greatly on the slopes of the demand and supply curves in the neighborhood of the equilibrium point B. It appears that if either the demand or the supply curve is very steep in this neighborhood, the dead loss will be slight. For a tax on the site rental value of land, whose supply curve is vertical, the dead loss drops to zero. A tax on site values is therefore one of the very best of all possible taxes from the standpoint of the maximum of the total national dividend. It is not difficult to substantiate this argument in dealing with related commodities; for the δq_i's corresponding to such a tax are zero. Since the incidence is on the owner of the land and cannot be shifted by any readjustment of production, it has the same advantages as an income tax from the standpoint of maximizing the national dividend. The fact that such a land tax cannot be shifted seems to account for the bitterness of the opposition to it. The proposition that there is no ethical objection to the confiscation of the site value of land by taxation, if and when the non-landowning classes can get the power to do so, has been ably defended by H. G. Brown.[10]

Land is the most obviously important, but not by any means the only good, whose quantity is nearly or quite unresponsive to changes in price, and which is not available in such quantities as to satisfy all demands. Holiday travel sometimes leads to such a demand for the use of railroad cars as to bring about excessive and uncomfortable crowding. If the total demand the year around is not sufficiently great to lead to the construction of enough more cars to relieve the crowding, the limited space in the existing cars acquires a rental value similar to that of land. Instead of selling tickets to the first in a queue, or selling so many as to bring about an excessive crowding that would neutralize the pleasure of the holiday, the economic way to handle this situation would be to charge a sufficiently high price to limit the demand. The revenue thus obtained, like the site value of land, may properly be taken by the state. The fact that it helps to fill the treasury from which funds are drawn to pay for replacement of the cars when they wear out, and to cover interest on their cost in the meantime, does not at all mean that any attempt should be made to equate the revenue from car-space rental to the cost of having the cars in existence.

Another thing of limited quantity for which the demand exceeds the supply is the attention of people. Attention is desired for a variety of commercial, political, and other purposes, and is obtained with the help of billboards, newspaper, radio, and other advertising. Expropriation of the attention of the general public and its commercial sale and exploitation constitute a lucrative business. From some aspects this business appears to be of a similar character to that of the medieval robber barons, and therefore to be an appropriate subject for prohibition by a state democratically controlled by those from whom their attention is stolen. But attention attracting of some kinds and in some degree is bound to persist; and where it does, it may appropriately be taxed as a utilization of a limited resource. Taxation of advertising on this basis would be in addition to any taxation imposed for the purpose of

---

[10] *The Theory of Earned and Unearned Incomes*, Columbia, Missouri, 1918.

diminishing its quantity with a view to restoring the property of attention to its rightful owners.

If for some reason of political expediency or civil disorders it is impossible to raise sufficient revenue by income and inheritance taxes, taxes on site values, and similar taxes which do not entail a dead loss of the kind just demonstrated, excise taxes may have to be resorted to. The problem then arises of so arranging the rates on the various commodities as to raise the required sum while making the total dead loss a minimum. A solution of this theoretical question, taking account of the interrelations among commodities, is given on p. 607* of the study of Edgeworth's taxation paradox previously referred to.

IV. Effect on Distribution of Wealth

We have seen that, if society should put into effect a system of sales at marginal cost, with overhead paid out of taxes on incomes, inheritances, and the site value of land, there would exist a possible system of compensations and collections such that everyone would be better off than before. As a practical matter, however, it can be argued in particular cases that such adjustments would not in fact be made; that the general well-being would be purchased at the expense of sacrifices by some; and that it is unjust that some should gain at the expense of others, even when the gain is great and the cost small. For example, it appears that the United States Government can by introducing cheap hydroelectric power into the Tennessee Valley raise the whole level of economic existence, and so of culture and intelligence, in that region, and that the benefits enjoyed by the local population will be such as to exceed greatly in money value the cost of the development, taking account of interest. But if the government demands for the electricity generated a price sufficiently high to repay the investment, or even the interest on it, the benefits will be reduced to an extent far exceeding the revenue thus obtained by the government. It is even possible that no system of rates can be found that will pay the interest on the investment; yet the benefits may at the same time greatly exceed this interest in value. It appears to be good public policy to make the investment, and to sell the electric energy at marginal cost, which is extremely small. But this will mean that the cost will have to be paid in part by residents of other parts of the country, in the form of higher income and inheritance taxes. Those who are insistent on avoiding a change in the distribution of wealth at all costs will object.

One answer to this objection is that the benefits from such a development are not by any means confined to the persons and the region most immediately affected. Cheap power leads, for example, to production of cheap nitrates, which cut down the farmers' costs even in distant regions, and may benefit city dwellers in other distant regions. A host of other industries brought into being by cheap hydroelectric power have similar effects in diffusing general well-being. There is also the benefit to persons who on account of the new

* *Editor's note*: Page 115 of the reprint in this volume.

industrial development find that they can better themselves by moving into the Tennessee Valley, or by investing their funds there. Furthermore, the nation at large has a stake in eradicating poverty, with its accompaniments of contagious diseases, crime, and political corruption, wherever these may occur.

A further answer to the objection that benefits may be paid for by those who do not receive them when such a development as that of the Tennessee Valley is undertaken is that no such enterprise stands alone. A government willing to undertake such an enterprise is, for the same reasons, ready to build other dams in other and widely scattered places, and to construct a great variety of public works. Each of these entails benefits which are diffused widely among all classes. A rough randomness in distribution should be ample to ensure such a distribution of benefits that most persons in every part of the country would be better off by reason of the program as a whole.

If new electric-power, railroad, highway, bridge, and other developments are widely undertaken at public expense, always on the basis of the criterion of maximizing total benefits, the geographical distribution of the benefits, and also the distribution among different occupational, racial, age, and sex groups, would seem pretty clearly to be such that every such large group would on the whole be benefited by the program. There are, however, two groups that might with some reason expect not to benefit. One of these consists of the very wealthy. Income and inheritance taxes are likely to be graduated in such a way that increases in government spending will be paid for, both directly and ultimately, by those possessed of great wealth, more than in the proportion that the number of such persons bears to the whole population. It would not be surprising if the benefits received by such persons as a result of the program of maximum total benefit should fall short of the cost to them.

The other class that might expect not to benefit from such a program consists of land speculators. If we consider, for example, a bridge, it is evident that the public as a whole must pay a certain cost of construction, whether the bridge be paid for by tolls or by taxes on the site value of land in the vicinity. There will be much more use of the bridge if there are no tolls, so that the public as a whole will get more for its money if it pays in the form of land taxes. But it will not, in general, be possible to devise a system of land taxes that will leave everyone, without exception, in a position as good as or better than as if the bridge had not been built and the taxes had not been levied. Landowners argue that the benefits of the bridge go to others, not to them; and even in cases in which land values have been heightened materially as a result of a new bridge, the landowners have been known to be vociferous in favor of a toll system. Payment for the bridge by tolls (when this is possible) has the advantage that no one seems to be injured, since each one who pays to cross the bridge has the option of not using it, and is in that case as well off as if the bridge did not exist. This reasoning is not strictly sound, since the bridge may have put out of business a ferry which for some users was more convenient and economical. Nevertheless, it retains enough cogency to stiffen the resistance of real-estate interests to the more economical system of paying for the bridge by land taxes.

Attempts at excessive accuracy in assessing costs of public enterprises according to benefits received tend strongly to reduce the total of those benefits, as in the case of the bridge. The welfare of all is promoted rather by a generous support of projects for communal spending in ways beneficial to the public at large, without attempting to recover from each enterprise its cost by charges for services rendered by that enterprise. The notion that public projects should be "self-liquidating," on which President Hoover based his inadequate program for combating the oncoming depression, while attractive to the wealthier taxpayers, is not consistent with the nation's getting the maximum of satisfactions for its expenditure.

V. DISTINCTION OF OPTIMUM FROM COMPETITIVE CONDITIONS

The idea that all will be for the best if only competition exists is a heritage from the economic theory of Adam Smith, built up at a time when agriculture was still the dominant economic activity. The typical agricultural situation is one of rising marginal costs. Free competition, of the type that has usually existed in agriculture, leads to sales at marginal cost, if we now abstract the effects of weather and other uncertainty, which are irrelevant to our problem. Since we have seen that sales at marginal cost are a condition of maximum general welfare, this situation is a satisfactory one so far as it goes. But the free competition associated with agriculture, or with unorganized labor, is not characteristic of enterprises such as railroads, electric-power plants, bridges, and heavy industry. It is true that a toll bridge may be in competition with other bridges and ferries; but it is a very different kind of competition, more in the nature of duopoly. To rely on such competition for the efficient conduct of an economic system is to use a theorem without observing that its premises do not apply. Free competition among toll-bridge owners, of the kind necessary to make the conclusion applicable, would require that each bridge be parallelled by an infinite number of others immediately adjacent to it, all the owners being permanently engaged in cut-throat competition. If the marginal cost of letting a vehicle go over a bridge is neglected, it is clear that under such conditions the tolls would quickly drop to zero and the owners would retire in disgust to allow anyone who pleased to cross free.

The efficient way to operate a bridge—and the same applies to a railroad or a factory, if we neglect the small cost of an additional unit of product or of transportation—is to make it free to the public, so long at least as the use of it does not increase to a state of overcrowding. A free bridge costs no more to construct than a toll bridge, and costs less to operate; but society, which must pay the cost in some way or other, gets far more benefit from the bridge if it is free, since in this case it will be more used. Charging a toll, however small, causes some people to waste time and money in going around by longer but cheaper ways, and prevents others from crossing. The higher the toll, the greater is the damage done in this way; to a first approximation, for small tolls, the damage is proportional to the square of the toll rate, as Dupuit showed. There is no such damage if the bridge is paid for by income, inheritance, and land taxes, or for example by a tax on the real estate benefited, with exemption of new improvements from taxation, so as not to interfere with the use of the land. The *distribution* of wealth among members of the community is affected by the mode of payment adopted for the bridge, but not the total wealth, except that it is diminished by bridge tolls and other similar forms of excise. This is such plain common sense that toll bridges have now largely disappeared from civilized communities. But New York City's bridge and tunnels across the Hudson are still operated on a toll basis, because of the pressure of real estate interests anxious to shift the tax burden to wayfarers, and the possibility of collecting considerable sums from persons who do not vote in the city.

If we ignore the interrelations of the services of a bridge with other goods, and also the slight wear and tear on the bridge due to its use, we may with Dupuit represent the demand for these services by a curve such as that in Figure 2. The total benefit from the bridge is then represented by the whole area enclosed between the demand curve and the axes, provided the bridge is free. All this benefit goes to users of the bridge. But if a toll is charged, of magnitude corresponding to the height of the horizontal line, the recipients of

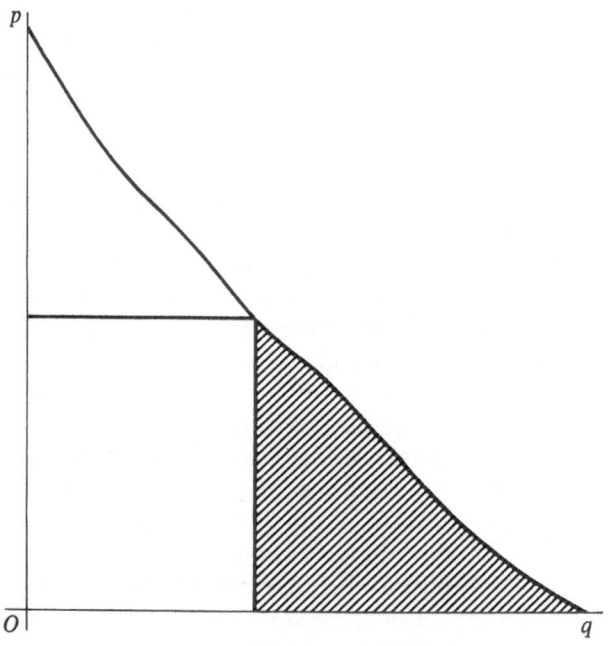

FIGURE 2

the toll are benefited to an extent represented by the area of the rectangle, whose base is the number of crossings and whose height is the charge for each crossing. But the number of crossings has diminished, the benefit to bridge users has shrunk to the small triangular area at the top, and the total benefit has decreased by the area of the shaded triangle at the right. This triangle represents the net loss to society due to the faulty method of paying for the bridge. If, for example, the demand curve is a straight line, and if the owners set the toll so as to bring them a maximum return, the net loss of benefit is 25 per cent of the total.

These are the pertinent considerations if the bridge is already in existence, or its construction definitely decided. But if we examine the general question of the circumstances in which bridges ought to be built, a further inefficiency is disclosed in the scheme of paying for bridges out of tolls. For society, it is beneficial to build the bridge if the total area in the figure exceeds the interest, amortization, and maintenance costs. But if the bridge must be paid for by tolls, it will not be built unless it is expected that these costs will be exceeded by the rectangular area alone. This area cannot, for our example of a linear demand function, be greater than half the total. We may in this case say that the toll system has 75 per cent efficiency in use, but only 50 per cent efficiency in providing new bridges. In each case the efficiency will be further diminished by reason of the cost of collecting and accounting for the tolls.

The argument about bridges applies equally to railroads, except that in the latter case there is some slight additional cost resulting from an extra passenger or an extra shipment of freight. My weight is such that when I ride on the train, more coal has to be burned in the locomotive, and I wear down the station platform by walking across it. What is more serious, I may help to overcrowd the train, diminishing the comfort of other travelers and helping to create a situation in which additional trains should be run, but often are not. The trivial nature of the extra costs of marginal use of the railroads has from the first been realized by the railroad managements themselves; indeed, it is implied in the amazingly complex rate structures they build up in the attempt to squeeze the last possible bit of revenue from freight and passenger traffic. If in a rational economic system the railroads were operated for the benefit of the people as a whole, it is plain that if people were to be induced by low rates to travel in one season rather than another, the season selected should be one in which travel would otherwise be light, leaving the cars nearly empty, and not a season in which they are normally overcrowded. Actually, our railroads run trains about the country in winter with few passengers, while crowding multitudes of travelers into their cars in summer. The rates are made high in winter, lower in summer, on the ground that the summer demand is more elastic than that of the winter travelers, who are usually on business rather than pleasure, and thus decide the question of a trip with less sensitiveness to the cost.

VI. COMPLEXITY OF ACTUAL RAILWAY RATES AND
REMOTENESS FROM MARGINAL COST

The extreme and uneconomic complexity of railway freight and passenger rate structures is seldom realized by those not closely in touch with them. A few random examples will illustrate the remoteness of actual rates from what may be presumed to be marginal costs, which railway managements will find it profitable to cover even by the lowest rates. Prior to the last enforced reduction of American passenger rates the regular round-trip fare between New York City and Wilkesbarre, Pa. was $11.04. But at various times between 1932 and 1935 round-trip tickets good for limited periods were sold at $2.50, $6.00, $6.10, and $6.15. Between New York and Chicago the roundtrip fare in the same period varied between $33 and $65 for identical accommodations. Between New York and Washington the ordinary round-trip fare was $18.00, but an "excursion rate" of $3.50 was applied spasmodically.

The lumber and logging activities of the country, which have been at a standstill for several years, are suffering from freight rates which in many important cases nearly equal, and even exceed, the mill price of the lumber. Thus from the large sawmills at and near Baker, Oregon, which produce lumber for the New York market, the freight amounts to $16.50 per thousand board feet. For No. 3 Common Ponderosa Pine, the grade shipped in largest quantities, the price of one-by-four inch boards ranged in the autumn of 1933 from $14.50 to $15.50 at the mill. Thus the New York wholesale buyer must pay more than double the mill price, solely on account of freight. The freight even to Chicago approximated the mill price. For No. 4 Common, also an important grade, the price was $12.50 per thousand board feet at the mill, but the New York buyer had to pay $29.00. A few months earlier, the prices were about $8 per thousand board feet less than those just given, so that the railroads received far more than the mill operators. It is hard to escape the conclusion that these high freight rates interfered seriously with the sale of lumber.

One advantage of the system of charging only marginal cost would be a great simplification of the rate structure. This is a great desideratum. It must not be assumed too readily that every purchaser distributes his budget accurately to obtain the maximum of satisfactions, or the most efficient methods of production, when the determination of the optimum requires the study of an encyclopedic railroad tariff, together with complicated trial-and-error calculations. Neither, from the standpoint of a railroad, can it be assumed that the enormously complex rate differentials have been determined at all accurately for the purposes for which they were designed. These complicated rate structures further contravene the public interest in that they enhance artificially the advantages of large over small concerns. When immense calculations are required to determine the optimum combinations of transportation with other factors of production, the large concerns are in a distinctly better position to carry out the calculations and obtain the needed information.

VII. MARGINAL COST DEPENDS ON EXTENT OF UNUSED CAPACITY

In the determination of marginal cost there are, to be sure, certain complications. When a train is completely filled, and has all the cars it can haul, the marginal cost of carrying an extra passenger is the cost of running another train. On the other hand, in the more normal situation in which the equipment does not carry more than a small part of its capacity load, the marginal cost is virtually nothing. To avoid a sharp increase in rates at the time the train is filled, an averaging process is needed in the computation of rates, based on the probability of having to run an extra train. Further, in cases in which the available equipment is actually used to capacity, and it is not feasible or is of doubtful wisdom to increase the amount of equipment, something in the nature of a rental charge for the use of the facilities should, as indicated above, be levied to discriminate among different users in such a way that those willing to pay the most, and therefore in accordance with the usual assumptions deriving the most benefit, would be the ones obtaining the limited facilities which many desire. This rental charge for equipment, which for passenger travel would largely take the place of fares, should never be so high as to limit travel to fewer persons than can comfortably be accommodated, except for unpredictable fluctuations. The proceeds from the charge could be added to the funds derived from income, inheritance, and land taxes, and used to pay a part of the overhead costs. But there should be no attempt to pay all the overhead from such rental charges alone.

Except in the most congested regions, there would, however, be no such charge for the use of track and stations until the volume of traffic comes to exceed enormously the current levels. An example is the great under-utilization of the expensive Pennsylvania Station, in New York City, whose capacity was demonstrated during the war by bringing into the city the trains of the Erie and the Baltimore and Ohio railroads. These trains are now required to stop on the New Jersey shore, constituting a wasteful nuisance which had existed before government operation, which was replaced by the more efficient procedure by the government, but which was resumed when the lines were handed back to their private owners.

VIII. THE ATTEMPT TO PAY FIXED COSTS FROM RATES AND PRICES CONTRIBUTES TO RIGIDITY AND SO TO INSTABILITY

One of the evil consequences of the attempt to pay overhead out of operating revenue is the instability which it contributes to the economic system as a whole. This is illustrated by the events leading to the depression. The immense and accelerating progress of science and technology led to the creation of new industries and the introduction of wonderfully efficient new methods. The savings from the new methods were so great that corporate profits and real incomes surged upwards. So large were the profits and so satisfactory the dividends that the operating officials of great industries did not feel under compulsion to push up the selling prices of their products to the levels corresponding to maximum monopoly profit. Because they kept their prices low, while paying relatively high wages, the physical volume of goods produced and transferred became enormous. The impulse to produce, with possibly some altruistic motives besides, tempered the desire for profits in many concerns. But under a profit system this could not last. As the prices of corporate shares rose, pressure developed to pay dividends equivalent to interest on the higher prices. This pressure would probably have led presently to gradual increases in the money prices of manufactured products, if the general level of prices had remained stationary. Such however was not the case. The general level of prices was declining.

And decline it must, according to the equation of exchange, when there was such a great new flood of goods to be sold. The vast increase in physical volume of goods, created by the new technology, called for a greater use of money, if the price level was to be maintained. This need was met for a time by increases in bank loans and deposits, and in the velocity of circulation. But neither bank loans nor velocities could continue to increase as fast as goods, and prices had to fall. The fall was not uniform. Corporations under increasing pressure to cover their overhead and pay high dividends out of earnings were strongly averse to reducing the selling prices of their products, when these selling prices were already below the points which would yield maximum profit. For several years prior to the crash, the prices of manufactured products stuck fast, while the proportion of national expenditure paid for these products continued to increase. This left a shrinking volume of money payments to be made for the remaining commodities, and these, including particularly the agricultural, had to come down in price. If, as the general price level fell, railroad, utility, and manufacturing concerns had reduced their selling prices proportionately, the prosperity of the years 1922 to 1928 might have continued. But such reductions in selling prices were not possible when an increasing volume of overhead charges had to be paid out of earnings. The intensified efforts to do this resulted in a pushing up of "real" prices of manufactured products—that is, of the ratios of their prices to the general price level—and of "real" transportation rates. Indeed, with a rapidly falling general price level, railroad freight rates, measured in money, were actually increased in 1931. This increase of 15 per cent on a large range of commodities, like the subsequent increases in suburban commuters' passenger rates, was obtained on the ground that the railroads needed the money to cover their overhead costs, though their operating costs had declined. Of course the effect was to make the depression worse, by stopping traffic which would have flowed at the lower rates. On the theory that bond interest and other such items must be paid out of operating revenues, the railroads were "entitled" to the higher rates, for their business had fallen off. But economic equilibrium calls for a rising rather than a declining supply curve; if demand falls off, the offer price must be reduced in order to have the offered services taken. This antithesis of rising railway rates, when general prices and the ability to pay are falling, well illustrates the disequilibrating consequences of the idea that overhead costs must be paid from operating revenues. There now seems to be a possibility of a repetition of the disastrous 15 per cent freight-rate increase in a time of decline.[11]

This explanation of the contrast of the prosperity of 1928 with the cessation of production in the following years rests upon the contrast of the system of prices which results from the whole-hearted devotion of different concerns to their own respective profits, with the system of prices best for the economic organism as a whole. Under free competition, with no overhead, these two systems of prices tend to become identical. Where there are overhead costs, competition of the ideally free type is not permanently possible. Monopoly prices develop; and a system of monopoly prices is not a system which can serve human needs with maximum advantage.

IX. CRITERION AS TO WHAT INVESTMENTS ARE SOCIALLY WORTH WHILE

When a decision whether or not to construct a railway is left to the profit motive of private investors, the criterion used is that the total revenue $\sum p_i q_i$, being the sum of the products of the rates for the various services by the quantities sold, shall exceed the sum of operating costs and carrying charges on the cost of the enterprise. If no one thinks that there will be a positive excess of revenue, the construction will not be undertaken. We have seen in Section V that this rule is, from the standpoint of the general welfare, excessively conservative. What, then, should society adopt to replace it?

A less conservative criterion than that of a sufficient revenue for total costs is that *if some distribution of the burden is possible such that everyone concerned is better off than without the new investment, then there is a prima facie case for making the investment.* This leaves aside the question whether such a distribution is *practicable.* It may often be good social policy to undertake new enterprises even though some persons are put in a worse position than before, provided that the benefits to others are sufficiently great and widespread. It is on this ground that new inventions are permitted to crowd out less efficient industries. To hold otherwise would be to take the side of the hand weavers who tried to wreck the power looms that threatened their employment. But the rule must not be applied too harshly. Where losses involve serious hardship to individuals, there must be compensation, or at least relief to cover subsistence. Where there are many improvements, the law of averages may be trusted to equalize the benefits to some extent, but never completely. It will

---

[11] Since this was written the Interstate Commerce Commission has allowed a part of this proposed increase and postponed consideration of a request for a passenger fare rise.

always be necessary to provide for those individuals upon whom progress inflicts special hardship; if it were not possible to do this, we should have to reconcile ourselves to greater delays in the progress of industrial efficiency.

Subject to this qualification of avoiding excessive hardship to individuals, we may adopt the criterion stated. In applying it there will be the problem of selecting a limited number of proposed investments, corresponding to the available capital, from among a larger number of possibilities. The optimum solution corresponds to application of our criterion to discriminate between each pair of combinations. The total amount of calculation and exercise of judgment required will not, however, be so great as might be suggested by the number of pairs of combinations, which is immense. Numerous means are available to shorten this labor. One of these is by the application of the line integral (3), namely

$$w = \int \sum h_i \, dq_i,$$

which provides a measure of value corresponding to the sum of consumers' and producers' surpluses. The part of w constituting the generalized consumers' surplus is (2); the validity of this line integral as a measure of an individual's increment of satisfaction corresponding to sufficiently small changes in the q's may be seen merely by replacing p_i in (13), by f_i, and noticing that for small changes the marginal utility of money λ changes little, so that f_i is very nearly proportional to the derivative of the utility function Φ. Hence the increment in Φ is proportional to the sum of the integrals of the f's, apart from terms of higher order; and the factor of proportionality λ is such as to measure this increment in money so as to be comparable to an increase in income. Similar considerations apply to the part of w corresponding to producers' surplus.

Defenders of the current theory that the overhead costs of an industry must be met out of the sale of its products or services hold that this is necessary in order to find out whether the creation of the industry was a wise social policy. Nothing could be more absurd. Whether it was wise for the government to subsidize and its backers to construct the Union Pacific Railroad after the Civil War is an interesting historical question which would make a good subject for a dissertation, but it would be better, if necessary, to leave it unsolved than to ruin the country the Union Pacific was designed to serve by charging enormous freight rates and claiming that their sum constitutes a measure of the value to the country of the investment. Such an experimental solution of a historical question is too costly. In addition, it is as likely as not to give the wrong answer. The sum of the freight and passenger rates received, minus operating costs, is not the line integral $w = \int \sum h_i \, dq_i$, which with some accuracy measures the value to society of the investment, but is more closely related to the misleading measure of value $\sum p_i q_i$. In other words, the revenue resembles the area of the rectangle in Figure 2, while the possible benefit corresponds to the much larger triangular area. The revenue is the thing that appeals to an investor bent on his own profit, but as a criterion of whether construction ought in the public interest to be undertaken, it is biased in the direction of being too conservative.

Regardless of their own history, the fact is that we now have the railroads, and in the main are likely to have them with us for a considerable time in the future. It will be better to operate the railroads for the benefit of living human beings, while letting dead men and dead investments rest quietly in their graves, and to establish a system of rates and services calculated to assure the most efficient operation. When the question arises of building new railroads, or new major industries of any kind, or of scrapping the old, we shall face, not a historical, but a mathematical and economic problem. The question then will be whether the aggregate of the generalized surpluses of the form (3) is likely to be great enough to cover the anticipated cost of the new investment. This will call for a study of demand and cost functions by economists, statisticians, and engineers, and perhaps for a certain amount of large-scale experimentation for the sake of gaining information about these functions. The amount of such experiment and research which could easily be paid for out of the savings resulting from operation of industry in the public interest is very large indeed. Perhaps this is the way in which we shall ultimately get the materials for a scientific economics.

Columbia University
New York, N.Y.

MANAGING EDITOR'S NOTE: Professor Frisch has written a brief criticism of Professor Hotelling's argument, but because of limitations of space it has had to be held over for publication in a later issue.—D. H. L.

The Relation of Prices to Marginal Costs in an Optimum System

In the July issue of *Econometrica*[1] I gave a new proof, taking account of the interrelations of commodities by methods not available in the times of Dupuit and Marshall, that in a specified sense maximum welfare requires that the quantity of each good consumed or produced by an individual shall be that corresponding to all sales being at marginal cost. This proposition has revolutionary implications, for example in electric-power and railway economics, in showing that society would do well to cut rates drastically and replace the revenue thus lost by subsidies derived largely from income and inheritance taxes and the site value of land.

Professor Frisch[2] correctly points out that these optimum quantities may in theory be achieved without the actual prices being equal to the marginal costs. It is enough that all prices be *proportional* to marginal costs; and this could theoretically be the case if every commodity and service were subjected to a tax proportional to its marginal cost. Professor Frisch also suggests the alternative (not seriously as a practical proposal, I think, but as a *curiosum*) that the optimum quantities be achieved by telling the individual that his income tax will be so adjusted that his total tax will be a certain fixed amount. This would be equivalent to deducting from income tax (not merely from taxable income) the other taxes paid. It would seem to have no advantage over simply obtaining all the revenue directly by the income tax and abolishing the excise taxes. And there is a practically insuperable difficulty in that the deductions to be made from income tax would have to be not the nominal taxes levied under the excise law, but the portions (sometimes positive and sometimes negative) falling on each individual. This raises the whole problem of the incidence of taxation on each commodity, which would have to be solved for each individual income-tax return. The incidence of taxation depends in turn on the demand and supply functions of all the commodities in the system, of which a knowledge sufficient for this purpose may be put down as unattainable. Any doubt of the appalling immensity of such a task will be

[1] "The General Welfare in Relation to Problems of Taxation and of Railway and Utility Rates," pp. 242–269.

[2] "The Dupuit Taxation Theorem," *Econometrica*, Vol. 7, pp. 145–150.

dispelled by a study of the attempts to determine demand functions statistically presented in the substantial and important new book of Henry Schultz.[3]

The practical difficulties of designing taxes proportional to marginal cost are also substantial. Something of an attempt in this direction is represented by general sales taxes. The experience of the American states and cities that have adopted these taxes during the last ten years as emergency depression measures has been far from happy. There is wholesale evasion through buying outside the jurisdiction of the taxing unit, through purchases less than the minimum amount taxable, and through barter. These taxes lend themselves to fraud on the part of retailers, who are designated as agents to collect the revenue from the public; and for a given amount of checking up to prevent fraud, the sales taxes are immensely more costly to collect than income taxes.

These sales taxes, and others that have been proposed, do not even satisfy the theoretical criteria with which we are concerned, for they do not tax services and are not proportional to the marginal cost to the seller. To achieve the end in view, sales taxes would have to be compensated by taxes at the same rate on wages, interest, rents, royalties, professional fees, theater admissions, and other items. The inclusion of wages, rents and interest would give such a tax something of the character of an income tax, but without the graduation by which an income tax takes account of the greater ability to pay of those with large incomes. Furthermore, to conform to the theoretical criteria, the tax must be proportional to *marginal* cost—not average cost, or price, as prices are at present determined; and there must be no other addition to marginal cost. The proposition stands that railways, power plants, and other industries with marginal costs far below the prices charged for their services, could be increased greatly in social efficiency by means of subsidies and reductions of rates. While their rates remain above marginal costs, general reductions of them benefit consumers—in terms of actual money—more than they cost the companies.

The money cost of the inefficiency of a system of excise taxes (of which some may be negative and thus constitute bounties of so much per unit) is definitely measurable in the same sense that the money cost of anything else is measurable. There is only the sort of ambiguity that arises from the effect on the general price level, and thus inheres in any question of the cost of a proposed expenditure; it may generally be disregarded when the cost is not too great in comparison with the total economy. For small excise taxes, if we are content with the assumption of linear functions in a small neighborhood of equilibrium, the approximation obtained on p. 254 [*Editor's note*: Here p. 152], namely

$$\tfrac{1}{2} \sum_i \delta p_i \, \delta q_i,$$

is sufficient. Here δp_i is the increment (positive or negative) in the price paid

[3] *The Theory and Measurement of Demand*, University of Chicago Press, 1938.

by a certain individual (or received by him) for a unit of the ith commodity, and $-\delta q_i$ is the increment in the quantity of this commodity that he consumes (positive) or produces (negative), as a result of the taxes. The foregoing sum over all commodities then represents the portion of the loss that may be imputed to this individual. The total loss is the sum of such expressions over all individuals. If t_i is the excise tax on the ith commodity and δQ_i the decrement in the quantity of this commodity resulting from the taxes, the total dead loss has the money value

$$\tfrac{1}{2} \sum_i t_i \, \delta Q_i,$$

apart from powers and products of the δQ_i, terms that will be unimportant for sufficiently small taxes. This sum is the direct generalization of the little shaded triangle in the classical diagram, Figure 1, p. 243 [*Editor's note*: Here p. 143], which has the vertical side t_i and the altitude perpendicular to this side equal to $|\delta Q_i|$. If excise taxes must for any reason be used, they should be designed to minimize it.[4] If by keeping all the taxes strictly proportional to prices, or in any other way, the δq_i can be made zero, it will follow that the δQ_i are zero, and the dead loss is zero.

The fact that the optimum q's are theoretically possible with prices proportional and not equal to marginal costs should not be construed as a defense, even in theory, of the attempt to cover the whole of railway costs, including interest on the original cost of construction, the president's salary, and the rusting of the rails, by levies on the traffic that passes. The system of rates that has evolved out of this attempt is grotesquely disproportionate to the marginal costs even of the railway services. Moreover even proportional additions to the marginal costs for all railway services, without additions in the same proportion to the marginal costs of highway, air, and water traffic, and all other costs, produces social inefficiency of the kind under discussion.

As Professor Frisch remarks, I should certainly not hold that *every* decrease of excise taxes is a good thing. If a toll charge is immutably fixed on a certain bridge, the abolition of a toll on another bridge across the same stream may or may not be a social gain. To decide this question requires a balancing of the losses on account of the traffic that does not cross because of the toll against the uneconomic travel that will be made to and from the free bridge by those who would otherwise find it more convenient to cross by the toll bridge, but go out of their way to avoid paying toll. In general, if a number of competing goods and services are originally taxed, and it is proposed to remove the tax from some but not all of them, the question whether the new or the old system is the worse cannot be decided without further data. We can only be sure that both are bad.

---

[4] Criteria for minimum dead loss consistent with a specified revenue have been discussed on pp. 256 and 257 of the July *Econometrica* paper, and from another point of view in my earlier paper on "Edgeworth's Taxation Paradox and the Nature of Demand and Supply Function," *Journal of Political Economy*, Vol. 40, 1932, particularly on p. 607. *Editor's note*: Here pp. 153–5 and p. 115, respectively.

We can, however, say that if we begin with an economy in which all sales are at marginal cost, the introduction of a system of excise taxes at varying rates per cent can never produce enough revenue to compensate for the loss of income tax that the state would have to endure if it were to undertake in this way to leave each person as well satisfied as before; provided we ignore the excessively improbable case of all the δq's being exactly zero. This was shown in my paper published in July. Conversely, the abolition of the excise taxes in such an economy will make it possible for each person to contribute more in other taxes to the state while retaining a position as satisfactory to himself as his previous one. And a *proportionate* reduction of all these excise taxes will conduce to the general welfare, though a special discriminatory reduction of one of them may not.

Some commentators on my paper have thought that something must be wrong because the proof that I gave of a system of taxes and bounties diminishing total satisfactions (in the special sense defined) might, it seemed, be reversed to show that after the imposition of the taxes and bounties, an abolition of them would further diminish total satisfactions. A suggestion of this is embodied in the fourteenth paragraph of Professor Frisch's remarks. The fallacy back of the paradox lies in the fact that the proof depends essentially on the original system being one of sales at marginal cost. After the excise taxes and bounties are imposed, sales are no longer at marginal cost; the theorem does not apply to this situation because its premise is not satisfied.

With reference to Professor Frisch's final paragraph, it may be pointed out that the expression $\frac{1}{2}\sum \delta p_i\, \delta q_i$ for an individual, and the expression $\frac{1}{2}\sum t_i\, \delta Q_i$ for the whole of the economy, represent correctly the social loss from a system of excise taxes in contrast to more efficient types, *regardless of what the government does with the money*. Whether the government spends the money wisely or badly, and the total amount to be spent by the government, are of course matters of grave public concern. But for a given total of taxation, these measures of the inefficiency of excise taxes, and other additions to marginal costs, are accurate so far as terms of the second order can go.

There is one relevant point which I have failed to bring out, but which is discussed by A. P. Lerner in somewhat similar connections. This is the fact that an income tax of the usual kind is a sort of excise tax on effort and on waiting, as well as on other less defensible ways of getting an income. An income tax is to some extent objectionable because it affects the choice between effort and leisure, and the choice between immediate and postponed consumption. Thus some of the same kind of loss attaches to an income tax as to excise taxes proper. How serious this effect may be is a question for factual research; but there is some reason to suppose an income tax superior to excise taxes on individual commodities in this respect, even apart from the objectives of progressive taxation and taxation of windfalls and other unearned incomes. To an extent depending on the degree of importance of this substitution of leisure for productive effort because of the income tax, and of immediate for postponed consumption, the public revenues, including those required to operate industries with sales at marginal costs, should avoid income taxes and should be derived primarily from rents of land and other scarce goods, inheritance and windfall taxes, and taxes designed to reduce socially harmful consumption.

Adam Smith, his predecessors, and his immediate successors were concerned with an economy consisting chiefly of agriculture, in which continually rising marginal costs with increasing output are the rule. In agriculture, moreover, there was free competition within national boundaries, with a resultant equality between price and marginal cost. The laissez-faire governmental policy advocated by classical economists was good and efficient under such conditions, apart from the question of the distribution of wealth. It is too frequently forgotten that the conditions of rising marginal cost and free competition are inconsistent with an industrial economy, and that consequently the classical arguments for laissez-faire have little application today outside of agriculture and small business. Our theorem includes all that is valid in the essence of the laissez-faire argument, which thus appears as a special case of the more general theory.

Columbia University
New York City

A Final Note

The proof of the fundamental theorems of my paper was on the basis of a set of assumptions including one to the effect that prices in the absence of the excise taxes were equal to marginal costs. It follows quite readily from these theorems that the attainment of the maximum of the general welfare, in the only sense that can apparently be given to this expression with the help only of rank ordering of a person's satisfactions without interpersonal comparisons, requires that all sales be at such prices that the quantities bought will be exactly the same as if these sales were at marginal cost. In particular, in considering what prices should be charged by state-owned or regulated enterprises, we have marginal cost as the criterion, leaving no place for the criteria generally used, such as average cost.

In the proof the change δr in government revenue from a person resulting from the imposition of a set of excise taxes, with a decrement δm in his income tax, was expressed (p. 250) [*Editor's note*: Here p. 148] by

$$\delta r = \sum q'_i \delta p_i - \delta m,$$

where δp_i is the increment in the price of the ith commodity resulting from the change, q'_i is the quantity this person buys of the ith commodity after the change, and $\sum$ denotes summation over all commodities. I showed that the change from income to excise taxes makes δr negative if the person's satisfaction is left unchanged; and that if, on the other hand, the rates are such that the revenue is unchanged, so that $\delta r = 0$, the person's satisfaction is diminished.

Now Professor Frisch's argument claims to apply to this a *reductio ad absurdum* by showing that any "self-financed" change, i.e., any change for which $\delta r = 0$, diminishes the person's satisfaction, even if prices were not originally equal to marginal costs. It is true, as both he and I have now proved, that every change making $\delta r = 0$, and not leaving all the quantities unchanged, reduces the person's satisfactions. His "entropic" interpretation seems to imply that changes in general make things go from bad to worse, that we had better endure present evils than fly to any other possible situation whatever, and that it is not only from a state of sales at marginal costs that changes are objectionable. He feels that in my proof no actual use has been made of the assumption of marginal cost being initially equal to price and that some undefined archaic system may be preferable to this social goal.

The key to this paradox lies in the interpretation of the condition $\delta r = 0$. This condition provides a suitable definition of a self-financed change *only if prices are originally equal to marginal costs*, or if the quantities sold are exactly as if prices were equal to marginal costs. If sales are originally at prices different from marginal costs, the differences constitute a revenue (positive or negative) whose magnitude will in general change as a result of the change in the price and tax system. Such a change in revenue must be considered in addition to the expression above for δr. It is not a matter of indifference to a government tobacco monopoly, for example, if the demand for tobacco is reduced by reason of a special tax on railway tickets required for riding in smoking compartments—unless the tobacco monopoly is selling at marginal cost.

Reverting to the case of a publicly owned toll bridge with which I began my original paper, let us consider that all costs resulting solely from use of the bridge are zero. The marginal cost of crossing is thus zero. Assume that a toll charge p (not regarded as a tax) is made per crossing, and that, at this rate, q crossings are made per year. There is thus a public revenue pq from this source. Whether this revenue is or is not required to pay interest on the cost of the bridge is not relevant to the present question. Now suppose that a tax t per crossing is levied on each user in addition to the toll; and that, because of the tax, the number of crossings per year thereafter is reduced to q'. In the foregoing expression for δr there is only one term $q't$ in the summation, in which δp_i is replaced by t and q'_i by q'. If we were to regard a measure as self-financed when $\delta r = 0$ and if the entire proceeds $q't$ of this tax were used to abate the income taxes of the persons crossing, we should have to call this a self-financed measure. The weakness of such a definition of self-financed measures is revealed by the fact that the government will not be in so good a position as before; for the reduction of the number of uses from q to q' will reduce its ordinary revenue pq to pq'. If we supposed that the combination of the tax on bridge crossings with reduction in income taxes really left the treasury in as good a position as before we should have to have $pq' = pq$. This would imply either that $q' = q$, i.e., that the increased cost of crossing had no effect on the traffic, or that $p = 0$, i.e. that the price charged for the use of the bridge was only the marginal cost. Since the possibility $q' = q$ is excessively improbable, we are forced to the conclusion that only when the original charge was the marginal cost is it justifiable to regard as self-financed those changes satisfying the condition $\delta r = 0$.

Likewise in general the condition $\delta r = 0$ has real significance only where the sales are in the first instance those corresponding to prices equal to the respective marginal costs. It is in this way that the argument depends essentially on the concept of marginal cost. It is necessary in demonstrating the loss of satisfactions to confine ourselves to changes that leave the total budget in balance—including revenues from tolls, postage, licenses, government-owned railways, etc.—and not merely the budget derived from revenues ordinarily called taxes.

It is in this way that the foregoing theorems, and the practical conclusions drawn from them in my paper regarding the efficiency of operations at marginal cost, have a significance that actually involves marginal cost in an essential way.

Columbia University
New York City

Income-Tax Revision as Proposed by Irving Fisher

What is ordinarily called income, and what the law calls income, may be divided into two parts, called savings and spendings. The exact distinction between these two is not so clear-cut as may at first appear, since spending is often for satisfactions spread over a considerable time in the future, and since savings yield a present satisfaction through the security, power, and independence they confer as well as the future satisfactions from the goods for which the savings may eventually be spent. Nevertheless the distinction can be made, somewhat roughly, when the satisfactions arising from the possession of savings are neglected, and when depreciation accounts are kept for such articles as family automobiles, furniture, and fur coats as well as for business property.

Professor Irving Fisher proposes to change the legal definition of income to mean consumption expenditures only, and to replace the present income tax by a tax on consumption. Such a change would avoid one serious inequity in the present arrangements, which results from the possibility of plowing profits into a business, or investing in land with a prospective increment in value equal to the market interest on an equally sound investment for the same time. Payment of income tax in such cases is postponed until the investment is "realized," with the result that the government loses the interest on the tax over what may be indefinitely many years. This evil might be dealt with by a periodical revaluation of assets, or by requiring that appreciation as well as depreciation be included each year in the calculation of income. The tax on undistributed corporate profits enacted a few years ago by Congress was a means of taking away the motive for stockholders to postpone paying their income taxes by keeping profits in the corporate treasury, where they could in effect be invested to earn further profits without the operation of formally passing into and out of the stockholders' hands, an operation that would require the payment of income tax. Repeal of the undistributed-profits tax has left wide open this avenue of evasion of the income tax. Professor Fisher's method of dealing with the problem is to remove the income tax from all that portion of income which is not spent, and to postpone the tax on the part whose spending is postponed. What the owner did not spend at all would not be taxed during his lifetime, but might be taxed at his death if it survived him.

Professor Fisher's theory of income is an internally consistent one, though there seems to be no good historical reason for applying the term "income" to expenditures. Nevertheless it does not take full account of the considerations of public policy that have led to the enactment of our present progressive income taxes. The rich miser who penuriously hoards his dividends, acquiring more and more control and power over the lives of others while living austerely, has a greater ability to pay than the poor workman who is forced to spend his whole earnings merely to keep on living and working. Yet the proposed change in the law would demand the same income tax from the workman as from the miser if their living expenses were the same. Neither the criterion of ability to pay nor that of equal sacrifice is satisfied by such a plan.

In a recent article[1] Professor Fisher undertakes to demonstrate the remarkable proposition that the government actually loses money by taxing the portion of income (in the ordinary sense) that goes into savings, instead of waiting until the owner's death and then taxing his estate at the same rate. Since this is true only in very special circumstances which may not be understood by non-mathematical readers, it is important that the question receive further elucidation. Professor Fisher's great and well-justified reputation may cause many people who cannot understand his mathematics to accept uncritically, without full consideration of the premises, conclusions which appear superficially to call for drastic modification and reduction of the income tax. On the other hand there may be some who, feeling the conclusion absurd, will be led mistakenly to criticize the use of mathematics in economics. Professor Fisher has done a service by putting the argument into mathematical form, since this makes it easy to identify the assumptions and compare them with actuality.

Under the heading "Fundamental Destructiveness," Professor Fisher first proves that a tax on savings, or capital-increase, when levied year by year, will reduce the capital formation during a sufficiently long period more than would a tax levied at the same rate on the accumulated savings once for all at the end of the period. He comments:

> This proposition is not listed as one of the paradoxes, since some people will probably not be surprised by it, though many will be surprised at the *magnitude* of the disparity in typical cases and also at the fact that the inequality is *reversed* when n is sufficiently small. [Italics his. Here n is the number of years of accumulation.]

---

[1] "Paradoxes in Taxing Savings," *Econometrica*, Vol. 10, April, 1942, pp. 147–158. This article refers also to an article in the magazine *Taxes* for August, 1941, in which Professor Fisher advanced the same conclusion for the guidance of the tax officials who read it without troubling them with the mathematical demonstration, which they would not be able to follow but which is published in the article in *Econometrica*.

A series of "paradoxes" is then demonstrated, including the propositions that
the taxpayer loses in capital more than the government gains in revenue from
a yearly tax; that less revenue is obtainable from yearly taxes on capital-
increase during life than from taxes (at the same rate) at death; that instead of
taxing the formation of an estate it would pay the government to borrow what
it thereby misses from its annual revenues and afterwards repay the loan out
of the tax on the estate; and that an increase in the yearly tax always defeats
itself in the end.

The qualification is made that the rate of interest paid by the government
must not be too high. The illustrative example given is of a hypothetical Henry
Ford who starts an automobile plant in 1900 with $1,000 capital, which grows
in value at 40 per cent compounded annually until 1940, when it amounts to
$700,500,000. At this time the hypothetical Mr. Ford dies, leaving the govern-
ment to tax his accumulation. But if the government had been so foolish as to
tax capital increments from the beginning at 20 per cent, the accumulation at
the time of death would have been only $66,500,000. On this the tax would
certainly be far less than a 20 per cent tax on the $700,500,000. Even taking
account of interest, the government would profit by postponing the tax—
*provided that interest is calculated at rates customary on United States govern-
ment securities*, such as two or three per cent. This proviso is essential. If the
government had to pay interest at forty per cent, compounded annually, to
parallel the portion of the profits of the Ford plant plowed into the business,
the case would be quite different.

Anyone who can find a really safe investment paying forty per cent per
annum, and who can borrow the necessary money at three per cent, is in a
fortunate position. Professor Fisher's argument merely shows that the government, like any private person, would do well to take advantage of such an
opportunity. His conclusions cannot be valid unless the government can
borrow money at a rate less than the rate of capital accumulation. The forty
per cent rate of the Ford example is sufficiently greater than the ordinary rates
paid on government securities to make this case seem clear. But can we be
sure that capital accumulation generally will be at a rate higher than the
government pays? Presumably uniform tax rates are to be applied, regardless
of the prospects of a business. Most businesses do not expand at anything like
the rate of forty per cent, compounded annually. Indeed, many businesses do
not expand at all. The long and sad record of business failures and dissipation
of capital may well put the government on guard against making indiscriminate
investments in the hope of garnering forty per cent interest on some of them.
Any banker would be cautious about lending at very high rates, or (what
comes to nearly the same thing) postponing payments due him, merely because
some concerns make handsome profits. Indeed, bankers have been so cautious
in this respect as to retard the early development of the automobile and other
industries. This partly explains the high rate of profit of some manufactures
who did manage to get capital. The proposal that the government forgo taxes
until the death of the owner implies that Uncle Sam should play the part of

a more generously openhanded banker, willing to give credit without the formality of inquiring into the soundness of the prospects of the individual borrower.

For the validity of the conclusion that the government will make money by waiting until death to tax capital gains, it is necessary that the rate of expansion of a person's capital shall exceed the interest rate on government securities over the period until his death, taken as a whole. If a man's capital expands for a few years, even at a rapid rate, and then remains stationary for a long period until his death, the conclusion may well be reversed. The compound interest on the forgone tax, even at two or three per cent, may amount to something very substantial if he lives a long time. If he loses his capital, whether through business ineptitude, the vanishing of markets, or any catastrophe, the government will lose both the tax and the interest on it. If in the last year of his life he chooses to squander his savings in one final carousal, the government will likewise lose completely both principal and interest of the tax it might have collected in earlier years, unless it has been foresighted enough to seize in advance the tax on these senile spendings which under the proposed change in the law would be the only trace of the income tax applicable to him at this time, or has transformed the vestigial income tax into a sales tax to be paid concurrently with the expenditures. Such a sales tax would have to cover *all* his expenditures, including wine, women, gambling, and hospital expenses of his last illness. The collection of such a tax might be very difficult. Thus the government in its capacity as banker would have to face all the usual risks of business failure, including the risk of embezzlement and defalcation. This last risk might be greatly enhanced both by the lack of selection among borrowers and by the sense of private property, which may cause many persons to feel a right to spend what is their own, without saving out money for the taxes which the government might wish to collect on the expenditures the following year under the name of income tax.

"Fundamental destructiveness" of a tax on capital gains is usually supposed by its opponents to result from the government squandering the money, so that the total amount of capital in the country is reduced. Professor Fisher avoids using this argument, and refrains explicitly from considering what the government does with the money. The discussions in recent years of the enormous capital value of government property have done much to take the edge away from the idea that the government destroys capital by taxation. A tax on capital gains may have no effect at all on the amount of capital in the country, but merely transfer a part of it to government ownership. In order to maintain that this is "fundamentally destructive," it is necessary to fall back on the widely popularized notion that government ownership of anything is inefficient. If this idea were carried to its logical conclusion it would hand over to private ownership not only the TVA but the highways and the rivers. It might lead the British government, as Bertrand Russell suggests, to turn over the job of defeating the German navy to the Thames Navigation Company. Professor Fisher is not a victim of these delusions. In the paper referred to, he rests the case against a tax on the portion of income saved upon a calculation designed to maximize government revenues, taking account of compound interest. It is important to notice that the premises underlying the calculation involve an assumption that capital is *certain* to expand at a rate greater than the interest rate on government securities; or else that the difference between the two rates is sufficient to take account of risk in such a way that the government would be well advised to invest in private enterprises, taken at random, in the hope that enough of them would copy Henry Ford's success to offset the losses and the interest.

Columbia University

Permissions Acknowledgements

Permissions from the University of Chicago Press for

J. Pol. Econ. 39 (1931), pgs. 137-75.
J. Pol. Econ. 40 (1932), pgs. 577-616.
J. Pol. Econ. 39 (1931), pgs. 107-9.
J. Pol. Econ. 49 (1941), pgs. 136-7.

Permission from Basil Blackwell for

Hotelling: "Stability in competition" *The Economic Journal*, 1929, 39, 41-57.

Permission from The Econometric Society for

1933. "Note on Edgeworth's taxation paradox and Professor Garver's additional condition on demand functions." *Econometrica*, 1, 408-9.
1935. "Demand functions with limited budgets." *Econometrica*, 3, 66-78.
1938. "The general welfare in relation to problems of taxation and of railway and utility rates." *Econometrica*, 6, 242-69.
1939. "The relation of prices to marginal cost in an optimum system." *Econometrica*, 7, 151-5.
1939. "A final note." *Econometrica*, 7, 158-60.
1940. "Appendix: questions of Preinreich," *Econometrica*, 8, 39-44.
1943. "Income-tax revision as proposed by Irving Fisher." *Econometrica*, 11, 83-7.
1939. "The work of Henry Schultz." *Econometrica*, 7, 97-103.

Permission from the American Statistical Association for

"A general mathmatical theory of depreciation" *Jour. of the Amer. Stat. Assoc.*, 1925, pp. 340-53.